Moore on Mercury

Sir Patrick Moore

Moore on Mercury

The Planet and the Missions

With 82 Illustrations

 Springer

Cover illustrations: Clockwise from bottom left: BepiColombo (image courtesy of ESA), Messenger (image courtesy of NASA) and Mariner 10 (image courtesy of the Jet Propulsion Laboratory). Background image of Mercury courtesy of NASA.

British Library Cataloguing in Publication Data
Moore, Patrick
  Moore on Mercury : the planet and the missions
  1. Space flight to Mercury 2. Mercury probes 3. Mercury
  (Planet)
  I. Title
  523.4′1
ISBN-13: 9781846282577
ISBN-10: 1846282578

Library of Congress Control Number: 2006931930

ISBN-10: 1-84628-257-8   Printed on acid-free paper
ISBN-13: 978-1-84628-257-7

9 8 7 6 5 4 3 2 1

Springer Science+Business Media
springer.com

Acknowledgements

My grateful thanks are due to Dr. David Rothery, of the Open University, for his invaluable help when this book reached the proof stage.

My thanks are also due to John Fletcher and Peter Paice for their photographs, and to Tony Wilmot for his drawings – and to NASA, JPL, the IAU and the Lowell Observatory for allowing me to use their material.

Patrick Moore
July 2006

Contents

1 Lift-off

In the early hours of August the third, 2004, a group of people waited expectantly at Cape Canaveral, in Florida. Most were scientists, though some were reporters and invited onlookers. All had eyes turned toward the launch pad, a safe distance away. In it was a tall rocket, carrying one of the most interesting space-craft of the new century: Messenger, bound for the little planet Mercury.

Many vehicles had been sent to other worlds; some had been successful, others not, and certainly there had been major developments since 1957, when Russia's Sputnik 1 ushered in the Space Age. Sputnik was no larger than a football, and carried little apart from the radio transmitter which sent out the "bleep! bleep!" signals which will never be forgotten by anybody who heard them (as I did). Since then almost all the attention of the space-planners had been focused on the Moon, Mars, Venus and the occasional comet, and only one probe had been sent to Mercury. This was Mariner 10, which had made three passes of the planet in 1973-74 before its transmitting power gave out. Mariner 10 had been a great success, but it had been able to image less than half of the planet's surface, and we had to admit that many problems remained unsolved, so that our in-depth knowledge of Mercury was decidedly sketchy. Messenger, it was hoped, would tell us much more.

The last moments before launch were tense. Lift-off had already been postponed once, mainly because of cloud-cover at Canaveral, but this time all went well. "Four … three … two … one … We have ignition!" There was

Figure 1.1. The launch of Messenger at Cape Canaveral on August 3, 2004. Images courtesy of NASA.

a brilliant burst of light, followed a few seconds later by a deafening roar as the rocket rose from its pad – slowly at first, then more and more rapidly. Messenger was on its way. Soon it was lost to the view of the waiting onlookers. Four minutes into the journey the first stage of the launcher fell away, and the motor of the second stage ignited to put the vehicle into orbit round the Earth. The onlookers breathed sighs of relief. After a 37-minute "coast" phase, the Delta-2 booster again fired for three minutes. There was a brief manœuvre, and the probe separated from the rocket; henceforth it

Figure 1.1. *Continued*

was on its own. The whole launch procedure had lasted for just under an hour. The two solar panels were deployed, to generate power, and the batteries were switched off. The five-thousand-million mile journey had begun.

Messenger – acronym *Mercury Surface Space Environment and Geochemistry and Ranging* – was in no particular hurry. Its Boeing-built Delta-2 launcher was not powerful enough to send the probe directly from Earth to Mercury, and it was necessary to use the gravity-assist technique, involving one swing-by of the Earth, two of Venus, and three of Mercury itself before the probe finally settles into Mercury orbit in March 2011. It is rather like driving from Brighton to Bognor Regis and making détours to Winchester and Hull, but there was no alternative, and the technique had already been tested and found to be satisfactory. Once in orbit, Messenger would spend well over a year collecting and transmitting data before being deliberately crash-landed on the surface of the planet in 2015 or 2016. This would not be a controlled landing, because this part of the original programme had to be abandoned on financial grounds. To President George W. Bush, Messenger was much less important than the attack on Iraq.

At least a great deal could be learned from observations made from orbit, and the space-planners were confident that the full programme could be carried through, even though it did mean rather a long wait.

Figure 1.2. The Earth from Messenger. Image courtesy of NASA.

2 Elusive Planet

ave you ever seen Mercury? Quite possibly the answer is "No", because you are not likely to glimpse it unless you are deliberately searching for it. When at its best, it is visible with the naked eye either very low in the west after sunset or very low in the east before dawn, but it always stays inconveniently close to the Sun in the sky, and can never be seen against a really dark background. Surprisingly, it can become brighter than any star, though Venus, Mars and Jupiter far outshine it. Stars are graded into magnitudes, according to their apparent brilliancy. The scale works rather in the way of a golfer's handicap, with the brighter performers having the lower magnitudes; thus magnitude 1 is brighter than 2, 2 brighter than 3 and so on. On a clear night, people with normal eyesight can see down to magnitude 6. Sirius, the brightest star in the sky, is of magnitude minus 1.5, and Mercury can achieve minus 1.9, but since it is seen against a bright sky it is easily overlooked.

It is the smallest of the planets in the Solar System (Pluto, alas, can no longer be classed as a planet). The diameter of Mercury is 3030 miles, which is not a great deal larger than the Moon (2158 miles) and much smaller than the Earth (7926 miles). The rocky surface is not good at reflecting sunlight, and Mercury is much further away than the Moon. The average distance of the Moon is 239,000 miles, but Mercury never comes much within 50 million miles of us, which is around two hundred times as remote as the Moon.

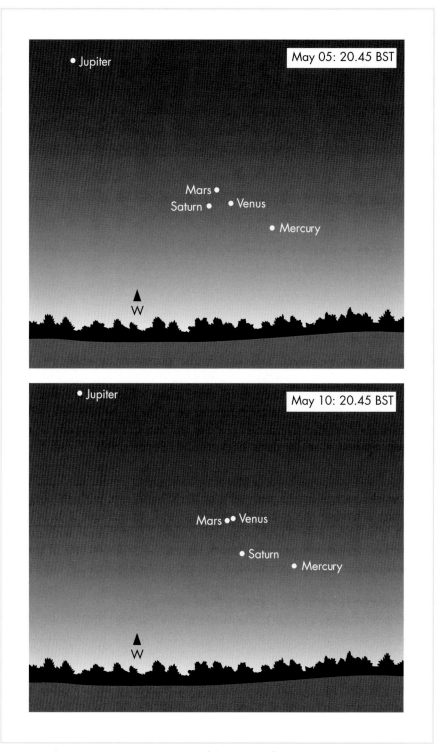

Figure 2.1. Positions of the planets five days apart.

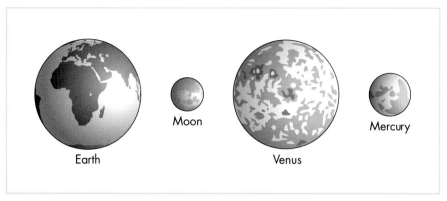

Figure 2.2. Relative sizes of the Earth, Moon, Venus and Mercury.

To set the scene, it may be helpful to say a little about the plan of the Solar System. Of course it is ruled by the Sun, which is an ordinary star. Astronomers classify it as a dwarf star, but it is vast compared with our puny world. Its diameter is 865,000 miles, so that it could swallow up well

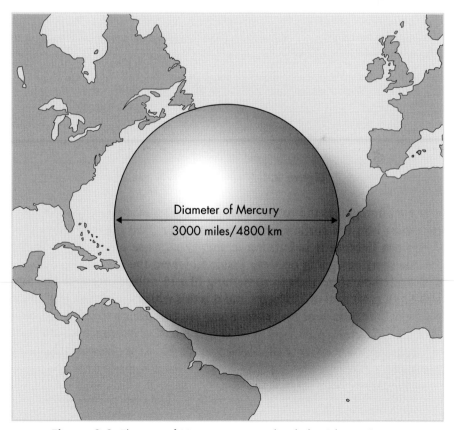

Figure 2.3. The size of Mercury compared with the Atlantic Ocean.

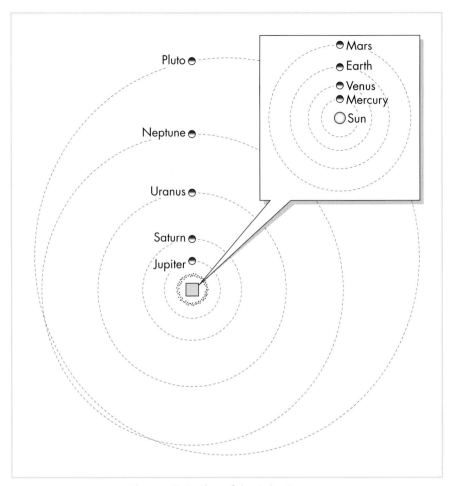

Figure 2.4. Plan of the Solar System.

over a million globes the volume of the Earth. It is made up of gas, and it is extremely hot; the surface temperature is over 5500 degrees Celsius.[1] It shines not because it is burning in the usual sense of the term, but by nuclear reactions going on deep inside it, where the pressure is colossal and the temperature rises to something like 15 million degrees. The Sun contains a great deal of hydrogen, and it is this which is used as "fuel". Near the Sun's core, nuclei of hydrogen atoms are combining to form nuclei of helium, the second lightest gas. It takes four nuclei of hydrogen to build one nucleus of helium; every time this happens, a little energy is set free and a

---

[1] Celsius temperatures are always used in astrophysics, and I have followed this, but for ordinary distance measures I have kept to Imperial units, so that everybody can understand them.

little mass (or weight, if you like) is lost. It is this energy which makes the Sun shine, and the mass-loss amounts to four million tons per second. This is not much relative to the total mass of the Sun, and there will be no really major changes in solar output for around a thousand million years yet. Eventually, of course, the source of available hydrogen will run low, and the

Figure 2.5. Relative sizes of the Sun and planets. From the top (most distant from the Sun): Pluto, Neptune, Uranus, Saturn, Jupiter, Mars, Earth, Venus, and Mercury. Illustration by Paul Doherty.

Sun will have to rearrange itself, with disastrous effects on the Earth, but there is no immediate cause for alarm!

Even a casual glance shows that the planetary system is divided into two distinct parts. First, there are four relatively small planets with solid surfaces: Mercury, Venus, the Earth and Mars. Beyond the orbit of Mars we come to tens of thousands of midget worlds, known variously as asteroids, planetoids or minor planets; only one (Ceres) is as much as 400 miles in diameter, and only one (Vesta) is ever visible with the naked eye. Then come the four giants Jupiter, Saturn, Uranus and Neptune, which have gaseous surfaces and relatively small silicate cores. Near and beyond the orbit of Neptune we come to another swarm of small worlds, making up the Kuiper Belt. Pluto, discovered in 1930, is the largest member of the swarm (diameter 1444 miles), but this is not great enough for Pluto to be classed as a proper planet. There are other Kuiper Belt objects, such as Quaoar and Varuna, with diameters more than half that of Pluto, and some are larger. There is one extraordinary object, Sedna – discovered in 2004 – which takes around 10,000 years to go once round the Sun, and goes out to an immense distance. It has even been suggested that Sedna has been "stolen" from another planetary system.

The four naked-eye planets have been known since prehistoric times, and Venus and Jupiter are particularly brilliant; so, at times, is Mars, while Saturn is always fairly conspicuous, and only Mercury tends to be elusive. Most of the planets have satellites. We have one – our familiar Moon; Mars has two tiny attendants, while Venus and Mercury are solitary wanderers in space, since any satellites even as much as a mile in diameter would certainly have been discovered by now. All the giant planets have satellite families, but of these only four bodies – three in Jupiter's system, one in Saturn's – are larger than our Moon. Pluto has three satellites – Charon, Nix and Hydra – but other Kuiper Belt objects, too, have attendants.

There may well be another large planet beyond the Kuiper Belt, but even if it exists – which is by no means certain – it is bound to be very remote and very faint. On the other hand, it was once believed that one planet orbited the Sun closer-in than Mercury, and it was even given a name: Vulcan. I will have more to say about it later.

It is true that Mercury is never striking, but if you look for it at times when it is best presented you ought to be able to see it on a couple of dozen mornings or evenings each year – provided, of course, that you live in the countryside, well away from glaring artificial lights; I fear that town dwellers will have no chance. For serious telescopic study it is best observed when high up, in broad daylight with the Sun above the horizon, which involves using a telescope equipped with setting circles. Sweeping around with an undriven telescope is emphatically not to be recommended, because it would be only too easy to look at the Sun by mistake, with tragic consequences for the observer. Permanent blindness could result, and, alas, accidents of this kind have happened in the past.

It cannot be said that a telescope will show much apart from the characteristic phase, but it is always satisfying to glimpse this strange little world shining coyly down from the brightness of the dawn or dusk sky.

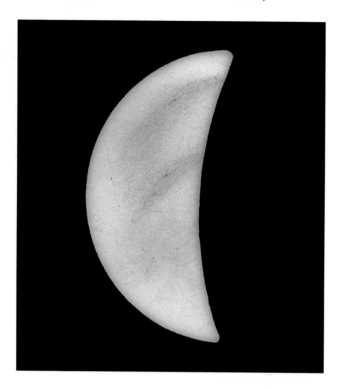

Figure 2.6. Drawing of Mercury by Patrick Moore, $12\frac{1}{2}$-inch reflector, 1956 April 13, 17h × 360.

3 "Messenger of the Gods"

I t is impossible to say who discovered Mercury. No doubt it was seen by our cave-dwelling ancestors, and it was certainly familiar to the first great civilizations, notably those of China and Egypt. The Egyptians named it Thoth, after the wisest of their gods; he is usually depicted as a man with the head of an ibis, though occasionally he becomes a dog-headed baboon (take your pick!). The earliest definite record dates back to 15 November 265BC, when according to Ptolemy, last of the great Greek astronomers, the planet was one lunar diameter away from a line joining the two bright stars Delta and Beta Scorpii.

Originally it was thought that the "evening star" and the "morning star" were two different bodies, but the Greeks soon realized that they were one and the same. They named it Hermes, after the messenger of the gods, because it was quick-moving, and was said to scurry from one side of the Sun to the other. When Rome became dominant, the Greek gods were given Roman names, so that Hermes became Mercury. (This is not quite the whole story, but it is a good approximation.) Roman mythology, too, is based upon that of Greece.

According to legend, Hermes was the son of Zeus (Roman, Jupiter) and the cloud-goddess Maia, who was also the goddess of spring, and the loveli-est of the Pleiades or Seven Sisters. You can see the Pleiades in the sky in the guise of a superb star-cluster, not far from the constellation of Orion, the Hunter. (Indeed, it was said that the Pleiades were placed in the sky by Zeus in order to put them out of reach of Orion, who was anxious to make their closer acquaintance but whose intentions were anything but honourable. At least Orion is a magnificent constellation, with his characteristic shape and his two outstanding stars, Betelgeux and Rigel, he dominates the night sky for a couple of months to either side of Christmas.)

Hermes – or Mercury – showed his initiative at an early age. When he was only a few minutes old he climbed out of his cradle and stole a herd of cattle belonging to Apollo, one of the major deities. Apollo was not pleased, but Mercury placated him by giving him a lyre, made from the shell of a tortoise, and teaching him how to play it. All was forgiven; Mercury was allowed to keep the cattle, and became one of the most popular of all the Olympians. He was a god who brought good fortune and wealth, and was associated with sleep and pleasant dreams. Yet he was also the patron of thieves and tricksters, so that his character was by no means "whiter than white". One has the feeling that in our modern world he would have made rather a good politician.

Figure 3.1. Hermes: the messenger of the gods. Image by courtesy of www.myastrologybook.com

Figure 3.2. Ptolemy.

We know rather less about him in other cultures, but he had various names; Enki to the Sumerians, Nabu or "the herald" in Assyria, Boudhavava in India (not to be confused with Buddha) and Woden in Persia. His "day" is Wednesday (Wodensdæg); in Latin Mercurii dies, in France Mercredi. Always he was associated with mobility and speed, and to the Chaldæans of Mesopotamia he was also a guide to the weather; if the planet shone brightly it meant either a very hot, dry summer, or a very cold winter, or both.

The astrologers did not neglect him, and said that his influence was dependent upon the circumstances, so that he differed from the benevolent Venus and Jupiter or the sinister Saturn and Mars. Remember, in ancient times astrology was regarded as a true science, and was still taken seriously up to the time of Newton.[2]

But how did the planets move? Here, religion came into the picture. One cornerstone of every ancient religion was that the Earth is the most important body in the universe, and that all other bodies revolve round it in perfectly circular orbits. The Greek philosophers soon realized that the

---

[2] For a scientific look at astrology, see my book *Stars of Destiny* (Canopus Publishing, Bristol 2006)

Earth is a globe rather than a flat plane, and a few of them (notably Aristarchus of Samos) went so far as to claim that it moved round the Sun, but no firm proofs were forthcoming, and the later philosophers kept to the old idea. What we know about ancient astronomy depends mainly upon one book, the *Syntaxis*, written around AD140 or 150 by Ptolemy (Claudius Ptolemæus) of Alexandria; the original has been lost, but the book has come down to us in its Arab translation, the *Almagest*. According to the Ptolemaic system, the Earth lies at rest in the centre of the universe; round it move the Moon, Mercury, Venus, the Sun, Mars, Jupiter, Saturn and the sphere carrying the "fixed stars".

Ptolemy was an excellent observer, and knew quite well that the planets do not move uniformly in circular orbits. To overcome this difficulty he maintained that each planet moved in a small circle or epicycle, the centre of which (the deferent) itself moved round the Earth in a perfect circle. As more and more irregularities were found, extra epicycles had to be introduced, until the whole system became very clumsy and artificial. Yet it did account for the movements of the planets, and few people dared to question it.

The real breakthrough was made by a Polish astronomer, Mikołaj Kopernik (Copernicus), who published his great book *De Revolutionibus Orbium Cœlestium* – Concerning the Revolutions of the Celestial Orbs – in 1543. He solved many of the main problems simply by removing the Earth

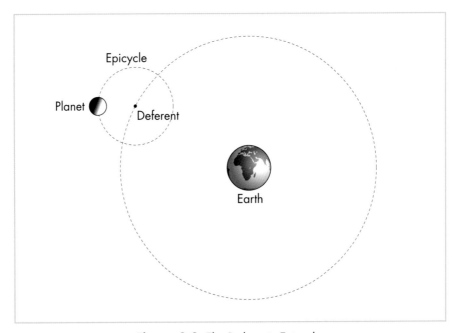

Figure 3.3. The Ptolemaic Epicycle.

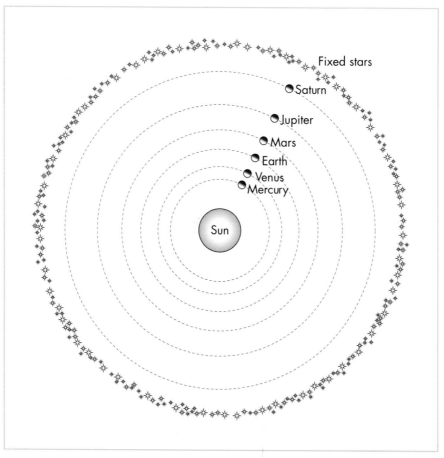

Figure 3.4. The Copernican System.

from its proud central position and putting the Sun there instead. True, he made many mistakes, and was even reduced to bringing back epicycles, but over the years more and more people adopted the Copernican system rather than the Ptolemaic, and the final proof came in 1687, when Isaac Newton produced his immortal *Principia*.[3]

Even before Newton, Mercury could have provided observational proof that it could not possibly move round the Earth. Telescopes had come into

---

[3] It is often said that Copernicus never saw Mercury at all, because of mists rising from the River Vistula, near his home. I am quite sure that this is wrong. In 1974 I was invited to give a lecture at Toruń University as part of the celebrations marking the quincentenary of Copernicus' birth. Mercury happened to be well placed at the time, and I stood just where Copernicus must have stood so long ago; I saw Mercury very clearly on eight consecutive evenings – and no doubt there is much more light pollution now than there was in 1473.

Figure 3.5. Copernicus University at Toruń. Photograph by Patrick Moore.

use early in the seventeenth century, and during the winter of 1609–10 the great Italian, Galileo, used his tiny telescope to make a series of spectacular discoveries. He saw the moons of Jupiter, the lunar mountains, the myriad stars of the Milky Way – and the phases of Venus. Venus shows phases in the same way as the Moon, from new to full, so that it can appear as a crescent, a half, or three-quarter (gibbous) shape before showing its complete disk. The Sun can light up only half of the Moon or a planet at any one time, and everything depends upon how much of the sunlit hemisphere is turned in our direction. The situation according to Ptolemaic theory is shown in Figure 3.6, where E represents the Earth, S is the Sun, and V1 to V4 shows Venus in four different positions in its epicycle. The line SVE must always be straight, and so Venus can never appear as a full or even a half-disk; it must always be a crescent, and this simple observation was enough to show that the Ptolemaic theory is wrong. Mercury too had been found to show phases, and the same principle would apply, but not much attention had been paid to Mercury, mainly because Venus is a much easier target.

So much for ancient times. Let us now come to our own era.

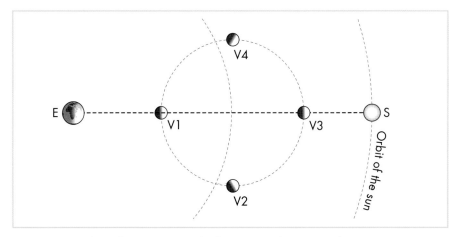

Figure 3.6. The Ptolemaic system. The line EDS is always straight; Mercury can never show a cycle of phases from new to full.

Mercury in the Solar System

4

ecause Mercury is the closest planet to the Sun, it is also the most rapid in its movements, which accounts for its name. The Earth moves round the Sun at an average speed of 18.6 miles per second; Mercury always travels much faster than that.

Way back in the early seventeenth century, the German mathematician Johannes Kepler worked out three famous Laws of Planetary Motion. Because they are so important, it will be useful to give them here:

1. A planet moves round the Sun in an elliptical orbit. The Sun occupies one focus of the ellipse, while the other focal point is empty.
2. The radius vector (the imaginary line joining the centre of the planet to the centre of the Sun) sweeps out equal areas in equal times.
3. The square of the orbital period is proportional to the cube of the mean distance from the Sun.

The first two Laws are self-explanatory. It is Law No 3 which needs a little explaining, so let me defer it for the moment.

All planetary orbits are elliptical. In an ellipse there are two foci, and the distance between them determines the eccentricity of the ellipse. If the foci coincide, the eccentricity is zero – in other words, a circle, so that a circle is merely an ellipse with an eccentricity of 0.

Consider the Earth. The eccentricity is 0.017, which does not differ much from a circle. At perihelion (closest approach) we are 91,400,000 miles from the Sun, while at aphelion (furthest out) the distance is 94,500,000 miles.

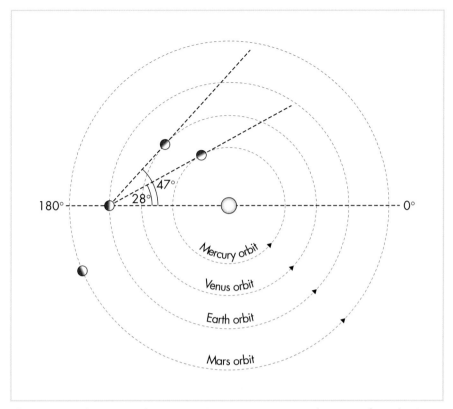

Figure 4.1. The inner Solar System. Mercury's maximum elongation from the Sun is 28 degrees; for Venus it is 47 degrees.

Some people are surprised to find that perihelion occurs in December, when it is winter in Britain, while the Earth reaches aphelion in June. The seasons are due entirely to the tilt of the Earth's axis of rotation, which is $23\frac{1}{2}$ degrees to the plane of orbit (more precisely, 23.439 degrees). In June the north pole of the Earth is tilted toward the Sun, so that in Britain we enjoy summer while it is winter in Australia; in December the situation is reversed. The fact that we are closer to the Sun in northern winter makes very little difference. In theory, southern summers should be hotter than those in the north, but this is masked by the stabilizing effect of the broad oceans which cover so much of the southern hemisphere.

The planets move in orbits which do not differ much from circles. The one real exception is Mercury, where the eccentricity is 0.205, and the distance from the Sun ranges between 28,600,000 miles at perihelion out to 43,500,000 miles at aphelion. This does make a marked difference; Mercury receives $2\frac{1}{2}$ times more heat at perihelion than it does at aphelion, but there are no Earth-type seasons, because the axial tilt of Mercury is below one degree to the orbital plane. In fact, it is almost "upright", which

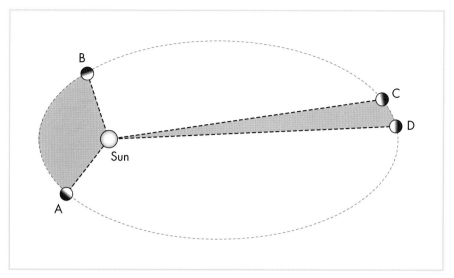

Figure 4.2. Kepler's Second Law.

cannot be said of any other planet in the Solar System, though it is true that for Jupiter the axial inclination is less than four degrees.

Now for Law No 2. As we have noted, the imaginary line joining the centre of the planet to the centre of the Sun is known as the radius vector, and Kepler used it in his "law of areas". In Figure 4.2, it is assumed that the planet moves from A to B in the same time it takes to move from C to D. (For the sake of clarity, I have drawn an orbit of high eccentricity, as is true of many comets – which of course obey the same Laws.) The shaded area ASB is equal to the shaded area CSD, and it is clear that the planet moves most rapidly when it is closest to the Sun. Mercury's orbital velocity

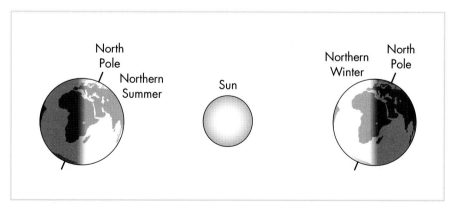

Figure 4.3. During a northern summer on Earth, the north pole is tilted toward the Sun.

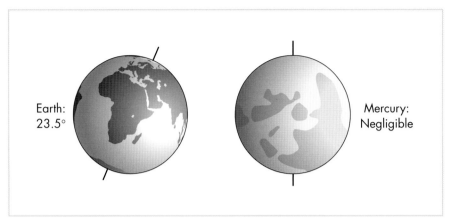

Figure 4.4. The axial tilt of Mercury compared with that of the Earth.

ranges from 35.4 miles per second at perihelion down to only 24.2 miles per second at aphelion. Neptune, the tortoise of the Sun's family, crawls along at an average speed of less than 4 miles per second.

Mercury and Venus, the only two planets closer to the Sun than we are, have their own way of behaving so far as we are concerned. They show phases, because they shine only because they are being lit up by the Sun, and obviously the Sun can shine upon only half of the globe at any one time, so that the opposite hemisphere is plunged in night. The boundary between the daylit and night hemispheres is known as the terminator.

Refer to Figure 4.5, which is not to scale. S represents the Sun, and Mercury is shown in four different positions in its orbit. At position 1 its dark face is turned toward us, so that Mercury is "new" and cannot be seen at all; this is known as the inferior conjunction. As it moves along in its path, the daylight side begins to turn towards us, and Mercury emerges into the dawn sky, first as a very thin crescent and then increasing to half-phase at position 2. It is then at its greatest angular distance from the Sun, though

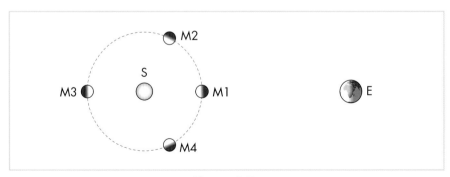

Figure 4.5.

greatest elongation can never be as much as 28 degrees. Mercury next becomes gibbous (between half and full), and then moves on to superior conjunction at position 3; it is full, but since it is practically behind the Sun it is to all intents and purposes unobservable. The phases are then repeated in the reverse order, with Mercury in the western sky after sunset: gibbous, dichotomy (position 4) and back to new position 1. The mean synodic period – that is to say, the interval between successive inferior conjunctions – is 115.9 days. Because of its quick motion, Mercury goes through its whole cycle of phases several times a year; thus in 2005 it was new on 29 March, 5 August and again on 24 November. A list of the phenomena for the period 2005-2012 is given in Appendix 2. When an "evening star", in the west, Mercury is waning, as is clear from the diagram; when a "morning star" at dawn, it is waxing. Of course we have to remember that the Earth's own movement round the Sun has to be borne in mind, but I have disregarded this in the diagram to avoid making it complicated. For the observer, Mercury has one particularly annoying trait. Its apparent diameter can range between 4.5 seconds of arc up to almost 13 seconds of arc, but the planet is closest to us when it is new, and cannot be seen. As the phase increases, so does the distance from the Earth, so that the apparent diameter shrinks; by dichotomy it is down to about 6 seconds of arc, which is not very much. When full, at superior conjunction, the apparent diameter is around 4 seconds of arc, but of course Mercury is out of view. All this, coupled with the small size of the planet, explains why we knew so little about the surface features before the flight of Mariner 10.

Turn now to Venus, which shows phases in the same way as Mercury, but is much easier to study because it is so much larger and closer. It can come to within 25 million miles of us, and far outshines any other body in the sky apart from the Sun and the Moon; at its best it can cast obvious shadows. (*En passant*, many people believed that Mars is the closest planet to the Earth, but this is not so; the minimum distance from Mars is over 34,000,000 miles.) There is one interesting difference between the phases of

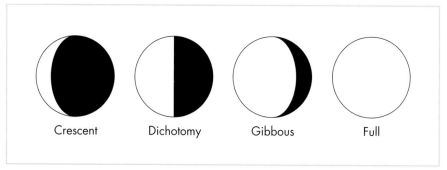

Figure 4.6. The phases of Mercury and Venus, as seen from Earth.

Figure 4.7. Schröter's drawings of Mercury.

Mercury and Venus. Since we have so perfect a knowledge of their orbits, we should be able to give an exact value for the phase at any one moment, but for Venus this does not work out. When an evening star, and therefore waning, dichotomy always occurs slightly ahead of the predicted time, while when waxing, in the morning sky, dichotomy is late. This discrepancy may amount to a few days, though of course it is never easy to give the exact time of half-phase, because Venus shows no surface features, and all we see is the top of the dense, cloud-laden atmosphere. The discrepancy was first noted in 1705 by the German astronomer Johann Schröter, and in a television broadcast I once referred to it as "the Schröter effect", a term which has now come into general use. It is certainly due to the dense atmosphere of Venus. There seems to be no comparable effect with Mercury, whose atmosphere is much too thin to make its presence felt. Over the past fifty years I have made many efforts to find a Mercurian "Schröter effect", with a total lack of success. Theory and observation always agree.

There is another difference, too. Venus is at its most brilliant during the crescent stage, while Mercury is brightest when gibbous. Again this is due to the fact that Venus has a dense atmosphere, whereas Mercury has not.

We are not sure who first recognized the phases of Mercury. They may have been suspected in the early seventeenth century by several observers using their primitive telescopes; they were definitely seen in 1639 by Giovanni Zupus, and they are within the range of a small modern instrument. The phase alters quite quickly, and it is fascinating to follow the sequence of events from one day to the next, even though you are unlikely to see any definite markings upon the tiny disk.

5 Crossing the Sun

I t is not quite true to say that Mercury can never be seen when 'new', at inferior conjunction. There are occasions when it is visible: when it is exactly lined up with the Sun, and appears in transit as a black dot against the brilliant solar disk.

If the orbits of the Earth and Mercury were in the same plane, a transit would occur at every inferior conjunction, but this is not so. Mercury's orbit is inclined to ours at an angle of 7 degrees, and this means that transits are relatively rare; everything has to be "just right". Moreover, transits can occur only in May and November, because of the ways in which the two planets move. The points where the orbit of Mercury crosses the ecliptic are known as the nodes, and a transit can happen only when Mercury is at or very near a node at the time of inferior conjunction. The actual dates are 7 May and 9 November. In November Mercury is closer to the Sun that it is in May, and as seen from Earth the apparent diameter of the Sun is greater, so that transits can occur as much as five days to either side of the 9th. For this reason November transits are over twice as frequent as those of May, but they do not last for so long. A list of transits for the period 2006–2050 is given in Appendix 2.

I had to wait many years for my first view of a transit of Mercury. Every time I was clouded out, or had no telescope available. I well remember that for the 1973 transit I was at Port Elizabeth Observatory in South Africa, where the weather is nearly always good at that time of the year. There was only one cloudy day during the whole of November – and that, of course,

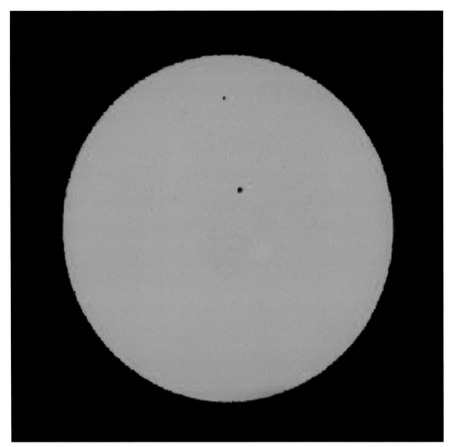

Figure 5.1a. Transit of Mercury in May 2003. Mercury is the tiny speck near the top of the disk; the large object near the centre is a sunspot. Orange filter. Photograph by John Fletcher, Gloucester, UK.

was the day of the transit. I finally succeeded on 7 May 2003. At Selsey the sky was brilliantly clear, the Sun was high in the sky, and I had a perfect view. Using my 5-inch refractor, I took a series of photographs (it is hardly necessary to add that my eyes were fully protected by special filters). It was a great moment for me when I first saw Mercury crawling slowly on to the solar disk.

Two things struck me at once. Of course I knew that during transit Mercury would be much too small to be seen with the naked eye, but I was surprised to find how tiny it was. I knew what the apparent angular diameter would be, but somehow I had expected Mercury to be more obvious than it actually was. Secondly, it was much blacker than a sunspot which happened to be on the Sun at the time. Spots are areas where the Sun's lines of magnetic force break through to the surface and cool it; the temperature of a spot is more than a thousand degrees lower than that of the surround-

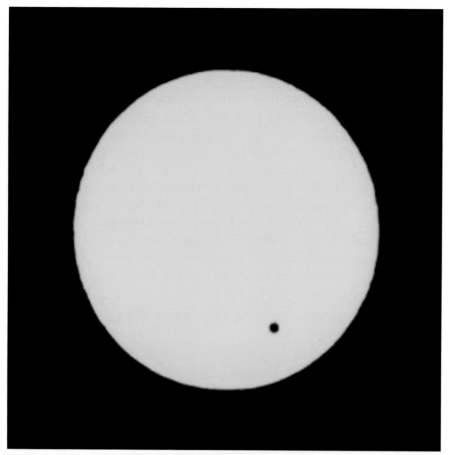

Figure 5.1b. Transit of Venus in June 2004, photographed by John Fletcher from Patrick Moore's observatory at Selsey, UK.

ing bright surface or photosphere, and it looks black only by contrast. If it could be seen shining on its own, its surface brightness would be greater than that of an arc-lamp.

The transit lasted for several hours, so that there was plenty of time to study it and to look for any unusual phenomena – very different from the frenetic activity during a total eclipse of the Sun! In fact nothing unexpected happened, and the strange events reported at earlier transits were absent. Almost certainly, these events can be dismissed as mere optical effects, but it is worth saying a little about them.

Reports are many and varied. Some refer to a bright or misty ring round the black disk, often attributed to a Mercurian atmosphere; Schröter said that he had seen white spots on the planet during transit, and these were also reported by other observers up to 1881, but not apparently after that. However, the "black drop" is one effect which I cannot discount (partly because I saw it myself at the transit of 2003), and this brings me on to a brief discussion of transits of Venus.

The orbital tilt of Venus is only just over 3 degrees, but transits are uncommon. They occur in pairs; a transit is followed by another transit eight years later, after which there are no more for over a century. Thus there were transits in 1631, 1639, 1761, 1769, 1874 and 1882; the last was in 2004, and the next will be in 2012, after which we must wait until 2117. During transit Venus, unlike Mercury, can easily be seen with the naked eye, always provided that proper precautions are taken. It, too, then seems blacker than a sunspot (though during the 2004 transit no major sunspots were on view). The first predictions of transits were made by Johannes Kepler. He said (correctly) that both Mercury and Venus would transit in 1631, Mercury on 7 November and Venus on 6 December. The Mercury transit was seen by four people, including a well-known French astronomer, Pierre Gassendi, and an equally well-known Swiss, Johann Baptist Cysat, usually known by his Latinized name of Cyastus. The others were

Transit of the Planet Mercury 7 May, 2003

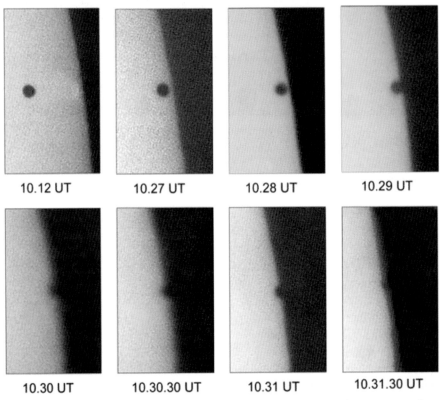

| 10.12 UT | 10.27 UT | 10.28 UT | 10.29 UT |

| 10.30 UT | 10.30.30 UT | 10.31 UT | 10.31.30 UT |

Figure 5.2. Transit of Mercury, May 7, 2003. Images by courtesy of Peter Paice, Belfast.

Figure 5.3. Edmond Halley.

Johann Remus Quietanus, about whom little seems to be known, but who lived in Alsace and corresponded with Kepler, and a Jesuit in Bavaria whose name has not been preserved. Prominent craters on the Moon have been named after Gassendi and Cysat, though Quietanus has no crater – history can sometimes be very unfair.

Nobody recorded the 1631 transit of Venus, because it took place during night over Europe, and at that time there were very few astronomers elsewhere. Kepler did not predict a Venus transit for 1639, but a young Englishman, Jeremiah Horrocks, did so, and actually managed to see it. (Horrocks seemed destined for a brilliant career, but, sadly, died when aged only twenty-one.) The other observer of that transit was one William Crabtree, who had been alerted by Horrocks.

At this point Edmond Halley comes into the story. To most people Halley is best remembered for his association with the famous comet which bears his name, and it is worth saying a little about this, because it was observations of a comet (not Halley's) which led to the first good determination of the mass of Mercury.

Comets are the most erratic members of the Solar System. They are not solid and rocky, like planets; a comet has a relatively small nucleus, made up of rocky particles embedded in ice, and often a tail or tails, made up of thin gas and tiny particles sent out from the nucleus. (I once described a comet as being "the nearest approach to nothing that can still be anything".) Many comets move round the Sun in periods of a few years, but most of them have orbits which are strikingly elliptical rather than nearly circular, and the comet can be seen only when moving in the inner part of the Solar System – though modern instruments can show some of them all round their paths. Originally it was not known that they had closed orbits; even Newton thought that a comet moved in a straight line, visiting the Sun only once before departing into the deep space from whence it came.

Halley was not so sure. In 1682 he observed a reasonably bright comet, and realized that it was moving in the same way as comets seen previously

Figure 5.4. Comet C/2000 C1 (Ikeya–Zhang) on April 16, 2002. Image by John Fletcher.

in 1607 and 1631. He deduced that these three comets were one and the same, and that the period was 76 years, so that the next return would take place in 1758. Halley knew that he would not live long enough to see it (he died in 1742) but, sure enough, the comet was discovered on Christmas Night 1758 by an astronomer named Palitzsch, and reached perihelion in 1759. The slight delay was due to the gravitational pulls of Jupiter and Saturn; a comet is of very low mass, and so is easily perturbed by a planet, even Mercury.

Quite apart from this, Halley made many important contributions to science, and it was also he who persuaded Isaac Newton to write the *Principia*; Halley even paid for its publication. But for the moment let us return to his connection with transits of Mercury.

In 1577 he was at the island of St. Helena, making the first careful catalogue of the far-southern stars which never rise over Europe. On 7 November he happened to observe Mercury passing across the face of the Sun. He was using "a good 24-foot telescope" (that is to say, a telescope with a focal length of 24 feet), and made very careful timings of the instants when the transit began and ended. It then occurred to him that a transit would, in theory, provide an excellent way of making one measurement which had always perplexed astronomers: the length of the astronomical unit, or distance between the Earth and Sun.

Up to then, nobody had any real idea of how far away the Sun was. Ptolemy, last of the Greek astronomers, had given an estimate of five million miles, which Copernicus reduced to just under 2,000,000 miles. Halley was sure that it was a great deal further away than that, but just how could he prove it?

Remember Kepler's Third Law, which gives a definite relationship between the orbital period of a planet and its mean distance from the Sun. The periods of the naked-eye planets were known very accurately, and so of course was the period of the Earth (just over 365 days), so it was a simple matter to draw up a scale map of the Solar System. What was needed was one absolute value – it did not matter for which planet, all the rest would then follow. Mercury could, in theory, give the vital clue.

Suppose that the transit was observed from two different places on Earth, a long way apart? Mercury would seem to cross the Sun in a different place to each observer. It would then be possible to work out the distance of Mercury, and everything else would fall neatly into place, including the length of the astronomical unit. In fact, it would be necessary to do no more than time the beginning and end of the transit from the two observing sites. So far, so good. But Halley also realized that Mercury's small apparent diameter would make it almost impossible to make the timings accurately enough. The answer was to make use of a transit of Venus.

The next transit was due in 1761 – and, as with the expected return of "his" comet, Halley knew that he could not hope to see it. So in a paper sent to the Royal Society in 1716, he wrote: "We therefore recommend again and again to our curious investigators of the stars … that they, mindful of our advice, apply themselves to the undertakings of these observations vigorously."

His advice was followed, and elaborate preparations were made. Unfortunately, one irritating phenomenon proved to be fatal to the whole method. As Venus drew on the Sun, it seemed to draw a strip of blackness after it – and when this strip disappeared, the transit had already begun. Nothing could be done about this "black drop" effect, as it was called, and it has to be said that the results from succeeding transits never came up to expectations. They are memorable in other ways, however; Captain Cook's voyage in 1769 was undertaken to take astronomers to Tahiti in order to observe the transit under favourable conditions, and it was only on his way back that he discovered Australia.

I was able to watch the last transit, that of 2004; it was well seen from my Selsey observatory, and over a hundred enthusiastic astronomers gathered there (it was quite a carnival atmosphere!). The Black Drop was very evident, and I managed to obtain a clear photograph of it. By then, of course, the mean distance of the Sun had been found in other ways, and given as 92,976,000 miles, but mainly for the sake of interest the distance was also derived from the transit, and worked out at 92,980,000 miles – an error of a mere 0.007 per cent.

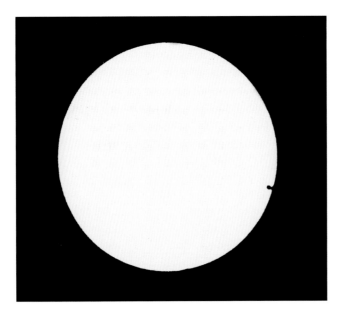

Figure 5.5. The Black Drop effect. Image by Patrick Moore, 2004.

With Venus, the Black Drop effect was originally blamed upon the dense, cloudy atmosphere, but this seems to be wrong; it is due to several factors, one being the falling-off of the brightness of the solar disk near the limb.

Now that the distance of the Sun has been so accurately measured, it must be admitted that transits of the inner planets have lost their importance, but they are still fascinating, and I hope to see the transit of Venus on 6 June 2012.

It is interesting, if not immediately useful, to consider transits of Mercury as seen from other planets, so here are some notes for the period between 2006 and 2020:

From Venus 4 June 2007, 3 June 2011, 18 December 2012, 17 December 2016. Obviously our observer will have to be above Venus' clouds.

From Mars 10 May 2013, 4 June 2014, 15 April 2015. The apparent diameter of Mercury will be about 6 seconds of arc.

From Jupiter 26 March 2006, 28 November 2011, 11 January 2018. Mercury's apparent diameter would be only 1.3 seconds of arc, but the transit would last for almost nine hours.

From Saturn 30 December 2011, 28 March 2012, 25 June 2012 and 22 September 2012. The apparent diameter of Mercury would be 0.75 of a second of arc, but the apparent diameter of the Sun would be only 3.5 minutes of arc, and a transit could last for up to 8 hours. 2012 will be a fruitful year so far as transits are concerned; if you happen to be holidaying on Saturn at that time, please remember to notify me so that I can set up a television link!

6 Ghost Planet

One of the leading French astronomers of the nineteenth century rejoiced in the name of Urbain Jean Joseph Le Verrier. About his brilliance there was no doubt at all; he was one of the best mathematicians of the time, and though primarily an astronomer he was also responsible for setting up the French meteorological survey. He is also said to have been one of the rudest men who had ever lived, and at one stage in his career he was dismissed as Director of the Paris Observatory because of his "irritability", though he was reinstated when his successor, Delaunay, was drowned in a freak boating accident. One of Le Verrier's colleagues said of him that "although he might not be the most detestable man in France, he was certainly the most detested".

I mention Le Verrier here because he was closely associated with studies of Mercury, and though his conclusions turned out to be wrong this was not his fault.

The story really began in 1781, when William Herschel, an astronomical enthusiast who was by profession an organist at the Octagon Chapel in Bath, discovered the planet we now call Uranus. Five planets had been known since antiquity – Mercury, Venus, Mars, Jupiter and Saturn – and nobody seriously thought that there might be another, if only because five planets, plus the Sun and the Moon, gave a grand total of seven, the mystical number of the ancients. Herschel was not looking for a planet, and did not even recognize it for what it was. He was interested in stars and nebulae, and was carrying out a systematic "review of the heavens", using

Figure 6.1. Urbain Jean Joseph Le Verrier.

a home-made telescope, when he came across an object which was not a star. It showed a disk, which no star can do, and it moved slowly from night to night against the starry background. Herschel thought that it must be a comet, but before long it was found to be a planet, moving far beyond the orbit of Saturn. It proved to be a giant, over 30,000 miles in diameter, with a gaseous surface; it takes 84 years to go once round the Sun, so that even now it has made less than three revolutions since its discovery. You can just see it with the naked eye on a clear night, provided that you know exactly where to look for it.

The mathematicians were soon busy, and worked out just how the new planet should move. This ought to have been fairly straightforward, but after a few years it became clear that something was wrong. Uranus was straying away from its predicted path – not by much, but by enough to be really puzzling. As time went by things became worse, and the mathematicians had to accept that some unforeseen factor was operating.

Of course, we know that each planet pulls upon the others (for example, the Earth's path round the Sun is not quite what it would be if Venus and the other planets were not there), but everything had been taken into account, and still Uranus obstinately refused to behave. There was nothing wrong with the observations; it was the calculations which were at fault.

Enter Urbain Jean Joseph Le Verrier. He looked carefully at the whole situation, and believed that he knew the cause of the trouble. If an unknown planet existed, moving at a greater distance from the Sun, it would tug upon Uranus and pull it out of position. Le Verrier decided that by working upon

Figure 6.2. William Herschel. Image by courtesy of The William Herschel Society.

the wanderings of Uranus, he could find out where the planet must be. It was rather like a cosmical detective story; he could see the victim – Uranus – and his task was to track down the culprit. By 1846 he was confident that he could give a definite position for it, and all that had to be done was to make a search with a suitable telescope.

Le Verrier was not himself an observer, at least at that time, and so he contacted the Paris Observatory, asking for a search to be put in hand. For some reason or other nothing was done quickly, and patience was never Le Verrier's strong point. His next step was to contact a German astronomer, Johann Galle of the Berlin Observatory, and ask for help.

Galle was enthusiastic, and took the matter to Johann Encke, Director of the Observatory. Encke was not particularly impressed, but felt that he could hardly refuse, so he commented "Let us oblige the gentleman from Paris". A good telescope was available, and Galle lost no time. Joined by a young member of the staff, Heinrich D'Arrest, he went into the dome as soon as darkness fell – and after only a few hours' work they found the planet. It looked rather like a star, but it did show a small disk, and the maps showed nothing in that position. Subsequently Encke contacted Le Verrier: "The planet whose existence you have predicted actually exists."

Figure 6.3. Johann Galle. Image by courtesy of the Deutsches Museum, München.

Figure 6.4. Johann Encke. Image by courtesy of Hamburg State Archives.

The name chosen for it was Neptune, after the mythological God of the Sea.[4]

That discovery made Le Verrier famous, and from a purely academic point of view his subsequent career was uniformly successful. His work extended into all branches of astronomy, but one problem he did not manage to solve; this related to Mercury.

As we have noted, Mercury's orbit is more eccentric than those of the other planets, and the distance of Mercury from the Sun ranges between 43 million miles at aphelion down to only 29 million miles at perihelion. Perihelion is reached every 88 Earth-days, and should fall in the same point of the orbit – but it does not; it moves forward in the orbit by a small but measurable amount at each circuit. As with the wanderings of Uranus, some external factor has to be involved, and several ideas were put forward during the nineteenth century. Could Mercury be moving through the outermost part of the Sun's atmosphere, so that it was being braked by friction? Or could the mass of Venus have been wrongly calculated, so that its pull on Mercury would be different from the accepted value? Or could

---

[4] It later emerged that calculations of the same kind had been made by a young Cambridge mathematician, John Couch Adams, and had given a very similar result, but no prompt search was made in England, and Neptune was not identified at Cambridge until after the announcement from Berlin. Many books credit Neptune's discovery jointly to Le Verrier and Adams, but this seems to be illogical; the actual discoverers were, without doubt, Galle and D'Arrest.

Figure 6.5. Heinrich D'Arrest. Image by courtesy of Wolfgang Steinicke.

there be another planet moving closer-in than Mercury, drowned in the brilliant glare of sunlight?

Mindful of his success with Uranus, Le Verrier not unnaturally favoured the third of these options, but it was obvious that tracking down a small body very close to the Sun would be a real problem. Probably the best hope would be to catch it when in transit against the solar disk, but even this would be far from easy; during transit Mercury looks very small indeed, and the suggested new planet would seem even smaller, even if it were the size of Mercury (which did not seem at all likely). Moreover, daily observations of the Sun had been made for many years, and no transit of an unknown body had ever been recorded. But before long there was an interesting development.

Le Verrier published his first paper about the motion of Mercury in 1859. Almost at once he was contacted by a French amateur astronomer, Dr Lescarbault, who claimed that he had observed such a transit in March of the same year. Lescarbault had his observatory at Orgères, in the Department of Euro-et-Loire, and Le Verrier made haste to go and see him.

It must have been a strange interview. Lescarbault was the village doctor, and was clearly overawed by his famous visitor, particularly after Le Verrier's opening gambit: "It is then you, Sir, who pretend to have observed the intra-Mercurian planet, and have committed the grave offence of keeping your observation secret for nine months. I warn you that I have come here with the intention of doing justice to your pretensions, and of demonstrating that you have been either dishonest or deceived." It was not a promising start, and it then emerged that Lescarbault had only a small telescope; that his timekeeper was a watch which had lost one of its hands, and that since he doubled as the local carpenter he used to record his observations in pencil on planks of wood, planing them off when he had no further need of them (paper must have been in short supply). Nothing could have been more bizarre.

In view of all this, it seems incredible that Le Verrier took the doctor seriously – but he did, and came away convinced that Lescarbault really had seen a planet in transit. He worked out a preliminary orbit, according to which the planet moved round the Sun at a mean distance of 13,082,000 miles in a period of 19 days 17 hours. The greatest possible elongation from the Sun, as seen from Earth, was a mere 8 degrees.

A name had to be found for it, and one seemed appropriate: Hephæstos, the blacksmith of the gods; this is the Greek name, Vulcan to the Romans. For a while, Vulcan found its way into many of the astronomical almanacs, particularly the French ones. But before long, serious doubts arose.

From Brazil, regular observations of the Sun were made by another French astronomer, Emmanuel Liais, who had a great deal of experience. He had been looking at the Sun at the exact time of Lescarbault's observation, and had seen nothing at all. He wrote: "I am in a position to deny, and deny positively and absolutely, the passage of a planet across the Sun at the time indicated." He went so far as to suggest that the whole story was pure fabrication. This may be because, in common with almost everybody else, he had an intense dislike of Le Verrier, but the facts could not be questioned.

Various other reports came in over the following years, but there was little doubt that all these related to ordinary sunspots (though one object seen in transit "appeared to be flapping", and was very reasonably attributed to a high-flying crane!). Le Verrier remained convinced of the true existence of Vulcan, but after his death in 1877 the whole matter might well have been forgotten but for some well-publicized observations made during the total solar eclipse of 29 July 1878.

During a total eclipse the sky darkens, and on most occasions planets and bright stars can be seen with the naked eye; even Mercury can be conspicuous. The 1878 eclipse was total over parts of North America, and two eminent observers, John C. Watson and Lewis Swift, independently set out to make a determined hunt for Vulcan. Their results were indeed surprising, since if their observations had been reliable they would have found not one Vulcan, but several. Yet careful analysis shows that in all probability they saw nothing more than two very ordinary stars, Theta and Zeta Cancri. Searches made at the next suitable eclipse, that of May 1883, were completely negative.

Two other episodes are worth mentioning here, if only because they introduce an element of farce. Many of the Vulcan believers put great faith in Lescarbault's transit report, but they may possibly have been worried by another Lescarbault observation, announced by him to the Académie des Sciences in January 1871. This related to an unidentified starlike object seen on 10 and 11 January of that year, not far from the star Regulus in Leo. He wrote: "I believe I saw it well, and was not the victim of an illusion." Kind friends were quick to point out that he had made a completely independent discovery of Saturn ….

Watson remained firm in his belief in the reality of Vulcan, and in 1879 he set out to prove his point by building a new and decidedly unusual observatory. He wrote: "I am erecting, at my own expense, a very elaborate observatory, one storey of which is underground and is to be used for special observations of the Sun and the inner planets, and for observations in the vicinity of the Sun." Remarkably, he seems to have been under the impression that stars can be seen in daylight if we observe from the bottom of a well or mine-shaft. This idea has been put forward time and time again, but a moment's thought will show how absurd it is. We cannot see stars in the daytime simply because of the lack of contrast between the starlight and the skylight, and observing from below ground does not make the slightest difference.[5] Nevertheless, Watson persisted. The shaft for his observatory was built into the side of a hill, and the observing room itself was simply a cellar from which a 55-foot pipe protruded. The pipe was aimed at the north pole of the sky, and at its top was a clock-driven mirror to direct the light from the target object down to a telescope installed in the cellar. Watson hoped that in this way it would be feasible to sweep close to the Sun and locate Vulcan. It was an arrangement worthy of Heath Robinson, but Watson collapsed and died before it was complete. A few tests were made by his successor, Edward Holden, but it was only too evident that the whole operation was a waste of time, and it was soon aban-

---

[5] I was once at the bottom of a deep mine-shaft in Poland, where in 1942 the Nazis had set up a concentration camp in a working mine. I had a good view of the sky, but certainly no stars came into view. The only unexpected phenomenon was a small black cat, which had made its home at the foot of the shaft where a restaurant had been installed for the benefit of tourists. The cat was a popular and pampered pet, and stubbornly refused to return to the great wide world above. Whether it is still there I know not, but at least it was very friendly and contented!

doned. The optical equipment was taken out, and eventually the "observa-tory" was filled in. It is safe to say that nothing like it has ever been built either before or since.

Not very much remains to be said. Vulcan refused to show itself at any eclipse, and the various reports of objects crossing the Sun were as un-convincing as Lescarbault's. Finally, in 1915, Albert Einstein published his General Theory of Relativity, and disposed of Vulcan once and for all. I do not propose to go very far down the "relativity road"; suffice to say that a massive body such as the Sun will distort space around it, and therefore affect the motion of any orbiting body. Actually this applies to all the planets, but only Mercury is small enough and close-in enough to make it evident. It was found that Einstein's theory accounted for the shift in the perihelion point of Mercury, and there was no need for Vulcan. So far as I know, the very last search was made at the total eclipse of May 1929. When this too was fruitless, Vulcan passed quietly into history.

However, we have not quite finished. Vulcan is a ghost, but we know some small members of the Solar System which invade the torrid region inside the orbit of Mercury. Many comets do so, but though comets may be huge (sometimes larger than the Sun itself) their masses are negligible, and could not possibly affect the movement of a planet. Neither can they be seen in transit. In 1910 it was found that Halley's Comet would pass right in front of the Sun, and a serious attempt was made to observe it, but though the transit time was known very precisely the comet was completely in-visible – and by cometary standards, Halley is large.

Asteroids are at least solid, and some of them stray far away from the main belt between the orbits of Mars and Jupiter. Here are various groups, each named after its leading member. Amor asteroids have orbits crossing that of Mars, but not that of the Earth; those of the Apollo group have orbits crossing those of both Earth and Mars; and Aten asteroids swing much closer-in, sometimes passing within the orbit of Mercury. (Phæthon was the first-known of these, and was appropriately named after the boy who, in mythology, was allowed to drive the Sun-chariot for one day, though unfortunately he lost control, almost set the world on fire, and had to be struck down by a thunderbolt.) However, all these are tiny, with diameters of only a few miles at most, so that they rank as cosmic débris. Some of them may well be ex-comets which have lost all their volatiles.

There are a few "Apohele" asteroids, whose paths lie entirely within that of the Earth (Apohele is Hawaiian for 'orbit'). It has been suggested that there may also be asteroids moving round the Sun in stable zones at distances of 7,440,000 miles and 19,500,000 miles – close to Le Verrier's value for Vulcan, so that these hypothetical asteroids have been termed Vulcanoids. Up to now no Vulcanoids have been found, despite ground-based searches and observations made by NASA using high-flying F-18 aircraft. This search was initiated by the Southwest Research Institute (SwRI) in collaboration with NASA's Dryden Flight Research Center at Edwards, California. During the three-flight observation campaign, two SwRI astronomers took a sophisticated digital system into the stratosphere in an F/A-18 jet aircraft. The altitude reached was 40,000 feet above the Mojave Desert. It is not likely that there can be any Vulcanoids with diameters of more than 40 miles or so, and even if they are of appreciable size and are brightly lit by sunlight, they will be very hard to detect against the overwhelming glare, but we know that the surface of Mercury is heavily cratered, and it is reasonable to suggest that the planet may have been hit by many Vulcanoids in the early story of the Solar System.

There, for the moment at least, the matter rests. Vulcan would have been an amazing and intriguing world, and it is rather sad to admit that it has proved to be no more than a wraith.

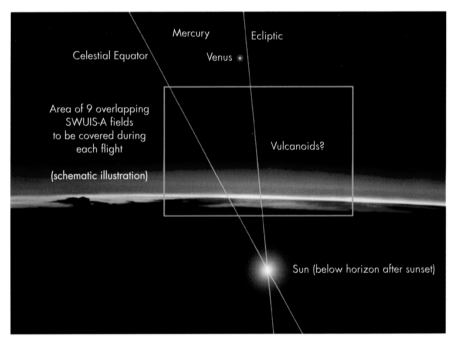

Figure 6.6. The search for Vulcanoids. Image courtesy of NASA.

Through the Telescope

Even small telescopes will show definite features on most of the bright planets. Mars has its polar caps, its ochre deserts and the dark areas once thought to be seas; Jupiter has its belts, its spots and also its satellites, while Saturn is distinguished by its glorious system of rings. Venus is less spectacular, because the actual surface is permanently hidden by the dense, cloudy atmosphere; there is no such thing as a sunny day on Venus. Mercury presents problems of a different kind, partly because it is small but mainly because it can never be seen against a really dark background.

However, a modest instrument will at least show the changing phases. These were probably suspected during the first part of the seventeenth century by a number of observers, including Galileo, and they were definitely recorded in 1639 by the Italian astronomer Giovanni Zupus. Surface details were much more elusive. William Herschel failed to see them during the late eighteenth century, even though his telescopes were much the best at the time, and the first published results were due to Johann Schröter, who had his observatory at Lilienthal, near Bremen in Germany, and was the real founder of selenography, the study of the surface of the Moon. (One of his telescopes was made by Herschel.) Whether the features shown on his drawings are genuine is by no means certain, though it must be noted that his drawings of Mars were remarkably good. The first serious attempts at mapping the Mercurian surface came considerably later.

In one way Mercury is less infuriating than Venus. There is no appreciable atmosphere, so that we can see right through to the surface. To say why this is so, I must bring in the question of escape velocity.

Throw an object upward and it will rise, stop and then fall back to the ground. Throw it faster, and it will rise higher before falling back. If you could throw it up at a speed of 7 miles per second (roughly 25,000 mph) it would never fall back at all; the Earth's gravity would be unable to hold it, and the object would escape into space. This is why 7 miles per second is known as the Earth's escape velocity.

Our air is made up of myriads of particles, all moving around. If a particle could move outward at 7 miles per second, it would be lost. Fortunately this cannot happen, because our air is made up of particles which can never work up to the critical speed, but the situation is very different for the Moon, which is much less massive than the Earth and has an escape velocity of only $1\frac{1}{2}$ miles per second. This is not great enough to hold down much in the way of atmosphere, and almost all of any air the Moon may once have had has leaked away. Today the Moon has only a trace left, and if we call it "an airless world" we are not very far wrong.

The other planets have escape velocities which depend upon their masses: 37 miles per second for Jupiter, 22 miles per second for Saturn, and so on, so that these giant worlds are surrounded by extensive atmospheres made up largely of hydrogen. Mars, with an escape velocity of just over 3 miles per second, does have an appreciable atmosphere, though it is much thinner than ours. Mercury – escape velocity 2.7 miles per second – is a borderline case, but its surviving atmosphere is so tenuous that it corresponds to what we normally call a laboratory vacuum. The total atmosphere seems to have a weight of only a few tens of tons, and has to be regarded as an exosphere; the particles collide with the surface of the planet more often than they do with each other. The ground pressure is so low that it certainly cannot support clouds of the kind with which we are so familiar.[6]

If you could go to Mercury – as no doubt men will do at some time in the future – you would find that you would weigh much less than you do at home, because the surface gravity is only 38 per cent of that of the Earth. If you weigh 12 stones on your bathroom scales, you would be reduced to only $4\frac{1}{2}$ stones on Mercury. On Mars you would also weigh $4\frac{1}{2}$ stones. Mars, diameter 4200 miles, is much larger than Mercury, but its mean density is only 4 times that of water, as against 5.5 times for the Earth and 5.4 times for

---

[6] Venus, with an escape velocity of over 6 miles per second, would be expected to have an atmosphere similar to ours, rather than the thick carbon-dioxide mantle which we actually find, but Venus is unusual in many ways, and though it is so alike the Earth in size and mass the two are certainly non-identical twins.

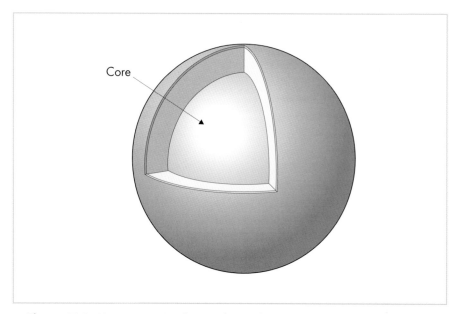

Figure 7.1. Mercury's iron-rich core, larger than the entire globe of the Moon.

Mercury. Mercury must have an iron-rich core which may be larger than the entire globe of the Moon; this is why the surface gravities of Mercury and Mars are the same. Lunar gravity is lower still, and this is clear from watching television pictures of men walking about there.

Not long ago I had a telephone call from a newcomer to astronomy who had just acquired a telescope – by amateur standards, a good one; it was a refractor with an object-glass 6 inches across. This will show a vast amount of detail on Jupiter and Mars, for example, but when he asked me what he would see on Mercury I was bound to reply, in all honesty, "Very little". The phase will be obvious, but the tiny disk, only a few seconds of arc in diameter, will not show any well-defined surface features. Drawings made with telescopes of this size are bound to be highly suspect, particularly when they show bright patches and dark streaks. Moreover, they seldom agree with each other. A really powerful telescope, used under good conditions, is essential.

When Mercury is visible with the naked eye it is bound to be low down, so that its light will come to us through a deep layer of the Earth's atmosphere, making the image unsteady. The only solution is to observe the planet when it is high in the sky. The Sun will also be visible, and this means that the observer will have to use a telescope fitted with accurate setting circles. Sweeping around Mercury in the daytime is most unwise; sooner or later the Sun will come into the field of view, with disastrous results.

Figure 7.2.
Schiaparelli.

The first observer to produce a useful map was Giovanni Schiaparelli, between 1881 and 1889; it was he who was the first to study Mercury in daylight. He was followed by Eugenios Antoniadi, whose map, produced in 1934, remained the best until the flight of Mariner 10, forty years later. Inexact though they were by modern standards, there is no doubt that these maps did show real markings on Mercury, but sadly, this cannot be said of the map produced by Percival Lowell, even though he was using one of the best and largest telescopes in the world, the 24-inch refractor at his observatory at Flagstaff in Arizona.

Lowell was in many ways a great man. He was a brilliant writer and speaker, a good organizer, and a first-rate mathematician; it was his work which led to the discovery of Pluto, in 1930 (even if this did owe something to luck). He came of a rich and aristocratic family, and became famous for his books and travels in the Far East. He could well have ended up as an eminent diplomat; instead he turned to astronomy, and concentrated upon the red planet Mars. Schiaparelli had used an excellent 8-inch refractor at Milan to produce surface maps, and had recorded straight, regularly-arranged lines which he called "canali"; this is Italian for "channels", but inevitably it was translated into English as "canals", and equally inevitably

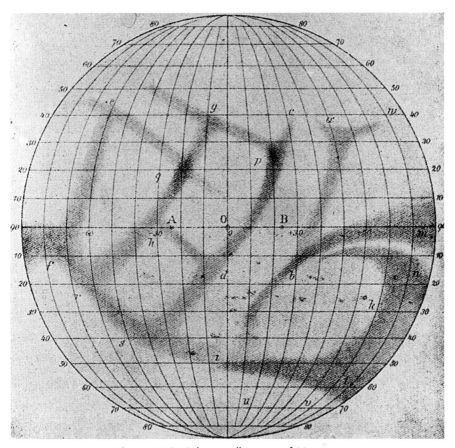

Figure 7.3. Schiaparelli's map of Mercury.

it was suggested that they might be artificial. Schiaparelli kept an open mind about this, but Lowell did not. He was convinced that intelligent Martians had built a planet-wide irrigation system, using canals to draw water from the polar ices through to the warmer regions at lower latitudes, and in 1908 he wrote: "That Mars is inhabited by beings of some sort or other is as certain as it is uncertain what those beings may be."

Never a man to do things by halves, Lowell established his observatory at Flagstaff and equipped it with one of the finest refractors ever made. (I know, because I made great use of it during my Moon-mapping days before the Apollo missions.) Lowell's one real weakness was that he was not a reliable observer, and he covered Mars with a network of canals which, if they had been real, could not possibly have been natural. We now know that they were mere tricks of the eye, though many other observers also "saw" them, and it was not until the Space Age that they were finally rejected. But Lowell also recorded linear features on other bodies – including Mercury.

Figure 7.4. Percival Lowell. Image by courtesy of the Lowell Observatory.

He wrote: "The lines on Mercury are far more difficult than the canals of Mars, for we see Mercury four times as far off when best placed as we do Mars. Though roughly linear, the streaks are not unnatural in appearance even at that great distance, and show irregularities suggestive of cracks."

Lowell never suggested that the features on Mercury might be artificial, but of their existence he had no doubt at all; it has to be said that his map shows no genuine features. He also recorded equally non-existent streaks on Venus, and was convinced that he was looking at a solid, rocky surface rather than a layer of cloud. It is a pity that he is so often remembered today for his "canals", because in other ways he carried out so much valuable work.

As an aside, I well remember my first view of Mars through Lowell's telescope. Conditions were good; I thought "Am I going to see canals?" I am delighted to say that I didn't. I also managed to make observations of Mercury, and I have to admit that I could see nothing on the disk apart from a few shadings.

Others, with better eyes than mine, fared better. But even they made mistakes which were not cleared up until a much later stage.

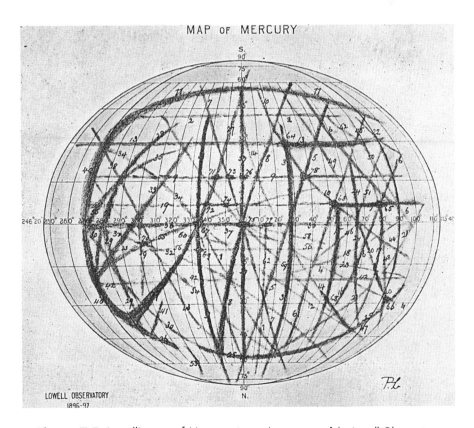

Figure 7.5. Lowell's map of Mercury. Image by courtesy of the Lowell Observatory.

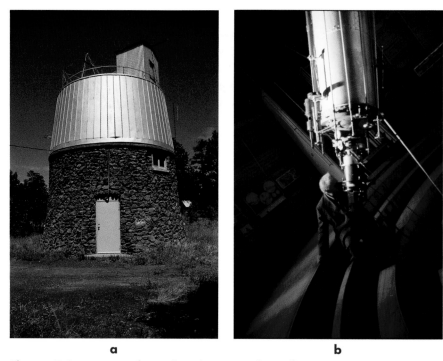

a   b

Figure 7.6. a Dome of Lowell's refractor at Flagstaff. Photograph by Patrick Moore.
b Patrick Moore with the Lowell 24-inch refractor.

8 Mapping Mercury

Many maps of Mercury were drawn up before the Space Age, mainly by amateurs but also by a few professionals. In general they did not agree well, but two were outstanding: those of Schiaparelli and Antoniadi.

I never knew Schiaparelli; he died in 1910, before I was born. He used the 8-inch and 12-inch refractors at the Brera Observatory in Milan. Though he recorded no "canals", he did draw features "which almost always showed up in the form of extremely delicate streaks, which under normal conditions could be recognized only with much difficulty and great concentration.... all these streaks are dark brown in colour against a rosy background, always smoky, not easily seen against the background, and difficult to distinguish". All in all he made a hundred and fifty drawings, but in 1889 he had to give up because of failing eyesight, and little more was done until Antoniadi came upon the scene.

I did know Antoniadi, who was Greek though he was actually born in Turkey; he went to France early in his career and spent the rest of his life there, taking French nationality. He died in Occupied France during the war, in 1944. For his main observational work he used the 33-inch refractor at the Meudon Observatory, outside Paris, which I know very well indeed. I had a few views of Mercury with it, but of course I was concerned mainly with the Moon, and I cannot say that I saw more on Mercury than I managed to do with Lowell's telescope at Flagstaff.

Figure 8.1. Dome of the 33-inch refractor at Meudon Observatory, used by Antoniadi. Photograph by Patrick Moore.

Antoniadi's map was quite detailed. It was published in 1934, together with a book dealing with all matters relating to Mercury, and became accepted as the standard work. For some curious reason no English version was available for many years. Eventually, in 1974, I translated it myself, though by then its interest was mainly historical.

The map showed bright and dark features, to which Antoniadi gave names; there was some agreement with Schiaparelli's chart, though there were also very marked differences. We now know that Mercury is a world of plains, mountains and craters; obviously Antoniadi could know nothing about these. He wrote that the patches were "very pale and difficult to distinguish, but they were genuine, and their colour seemed greyish, similar to that of the lunar seas". They included Solitudo Hermae Trismegisti (Wilderness of Hermes the Thrice Greatest), "a vast shading; it is comparable in area with Australia, and is the largest of the dark patches on the disk". There was Solitudo Criophori, "a very dark feature, 1550 miles long, curved with its point or horn diverging toward the north-east"; also Solitudo Atlantis, " a very dark patch in the south-west quadrant", which corresponds to a similar patch on Schiaparelli's chart, and was "approximately circular, though its closeness to the limb makes it appear elliptical; its greatest diameter is about 780 miles". Bright areas included Argyritis, composed of "a small, extremely brilliant spot, situated in the middle of a large, extended bright region"; Antoniadi added that it had been discovered in 1872 by a well-known English observer, W. F.

Denning, with a $12\frac{1}{2}$-inch reflector. Solitudo Jovis was "a rounded patch, very dark; dimensions very nearly equal to those of France… it was nearly always easy to see with the great refractor, and very dark when not enfeebled by local veils".

The reference to "local veils" is very interesting. Antoniadi believed firmly that Mercury had an atmosphere dense enough to support clouds of fine dust, and he reported local clouds which were "more frequent and obscuring than those of Mars". For once he was wrong, because, as we have noted, the Mercurian atmosphere is absolutely negligible; there is no chance that it could make its presence felt in any way. Later Audouin Dollfus at the Pic du Midi Observatory in the Pyrenees suspected traces of an atmosphere, but there was no confirmation. One remarkable experiment

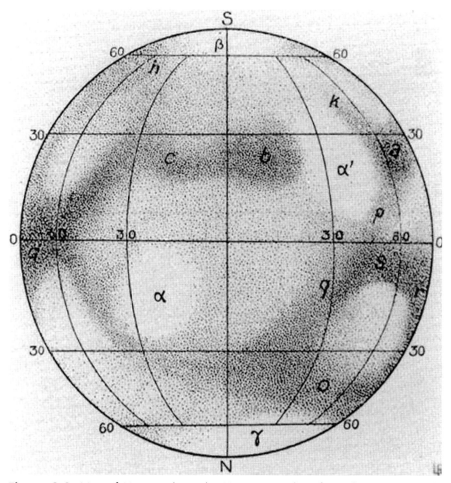

Figure 8.2. Map of Mercury drawn by M. Lucien Rudaux form observations made from 1893 to 1927 at the Donville (Manche) Observatory. Image courtesy of NASA.

Figure 8.3.
Antoniadi.

was carried out in 1961 by the Russian astronomer N. A. Kozyrev, at the Crimean Astrophysical Observatory. On February 15 there was a total eclipse of the Sun, and Kozyrev did not look at it at all; throughout the brief totality he made spectroscopic observations of Mercury, for once seen against a tolerably dark sky. He hoped to detect traces of atmosphere, but he had no success.

There can be no dusty veil over Mercury. In view of Antoniadi's skill and experience, it is hard to fathom how he could have been so wrong on so important a matter, but there is no other explanation. His other major mistake – concerning the planet's rotation period – was understandable, and under the circumstances probably unavoidable.

Schröter, the first really serious observer of Mercury, believed the axial rotation period to be just over 24 hours, much the same as ours. Denning, in 1882, preferred 25 hours. It is easy enough to measure the rotation period of Mars, because we can watch the surface features being carried across the disk, but this is not so easy with Mercury. Schiaparelli made a careful series

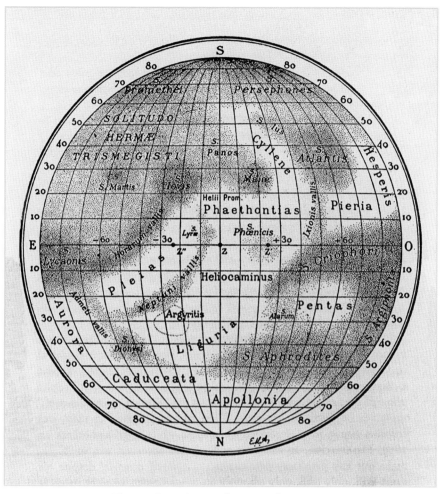

Figure 8.4. Antoniadi's map of Mercury.

of observations, and found that the shadings he could make out always appeared in the same positions on the disk. From this he concluded that the axial rotation period was the same as the orbital period; 88 Earth-days, so that a Mercurian "day" would be equal to a Mercurian "year". The rotation would be "captured", or synchronous.

There is an example of this much nearer home, with the Earth-Moon system. The Moon has an orbital period of 27.3 days, and its axial rotation period is exactly the same. Neglecting some minor effects known as librations, this means that the Moon keeps the same face turned Earthwards all the time.[7] There is no mystery about this; tidal friction over the ages has been responsible. Originally the Moon rotated much more quickly, but the Earth's tidal pull slowed it down until relative to the Earth (though not relative to the Sun) rotation had stopped. If Mercury behaved in the same way with respect to the Sun, as Schiaparelli believed, there would be everlasting day over one hemisphere of the planet and permanent night over the other, with only a narrow "twilight zone" in between – where the Sun would bob slowly up and down over the horizon.

Antoniadi was in full agreement with Schiaparelli, and the 88-day rotation period became so well established that nobody felt inclined to question it. Moreover, it seemed to be the only value which would fit the maps. But then, in 1962, came some disquieting news. At Michigan, in America, W. E. Howard and his colleagues measured the long-wavelength radiations coming from Mercury, and found that the dark side was much warmer than it would be if it were always turned away from the Sun. Something was badly wrong.

At this juncture, radar astronomy was called in. Basically, radar works by transmitting a pulse of radio energy, bouncing it off a solid body (or something which acts in a similar way), and then recording the "echo". Mercury may be a small, elusive target, but by the mid-1960s it was well within radar range. When pulses are reflected from a rotating body, the "echo" is affected, and the rate of spin can be found. This was done for Mercury, initially by Rolf Dyce and Gordon Pettingill, using sensitive equipment at Arecibo in Puerto Rico. They found that the rotation period must be much less than 88 days, and the modern value is 58.6461 days – two-thirds of a Mercurian year. Schiaparelli and Antoniadi had been misled by a very unexpected set of circumstances.

I do not want to delve into mathematics, even very simple ones, so I will do my best to sum up matters as concisely as possible, and give the overall results. The facts are as follows:

1. The synodic period of Mercury – that is to say, the time which elapses between successive appearances at the same phase – is, on average, 116 Earth days. If Mercury is "new" on a particular date, it will again be "new" 116 days later, and so on.

---

[7] If you want a simple demonstration, put a chair in the middle of the room and then walk round it, turning so as to keep your face turned toward the chair all the time. When you have completed the circuit, you will have turned on your axis because you will have faced every wall of the room, but anyone sitting on the chair will never have seen the back of your neck. Similarly, from Earth we never see the "back" of the Moon, and before the flight of the first circum-lunar probe, in 1959, we had no direct information about it. Note that day and night conditions are the same over both hemispheres of the Moon.

2. The axial rotation period (58.64 days) is equal to two-thirds of the orbital period (88 days). The axial inclination is less than one degree, so that with reference to its orbital plane Mercury is "upright".

3. It follows that to an observer at a fixed position on Mercury, the interval between one sunrise and the next will be 176 Earth-days, or 2 Mercurian years.

4. This interval, 176 Earth days, is approximately equal to $1\frac{1}{2}$ synodic periods.

5. From this, it is found that after every 3 synodic periods, the same face of Mercury will be seen at the same phase.

6. Three synodic periods of Mercury add up to one Earth year. Consequently, the most favourable times for looking at Mercury recur every three synodic periods. Glance back at Point 5. You will realize that every time Mercury is best placed for observation, we see the same hemisphere, with the same markings at the same positions on the disk.

Antoniadi made the best use of his opportunities, but even with the great Meudon refractor the surface features are hard to make out. He could see them best every third synodic period – and each time, he saw the same markings: in fact, those drawn on his map. He could not be expected to look for so curious a relationship, and it was natural for him to think that the rotation period must be synchronous. There was nothing wrong with the observations themselves; it was the interpretation which was faulty. True, the relationship is not exact, but it was good enough to deceive all observers for many years.

We are now in a position to work out the weird calendar of Mercury, which is unlike that of any other planet in the Solar System, and is far from being straightforward.

Because Mercury's orbit is decidedly eccentric, the orbital speed ranges from 24 miles per second at aphelion up to 36 miles per second at perihelion, but the rate of axial rotating spin remains constant. Near perihelion, the orbital angular velocity exceeds the constant spin angular velocity, so that an observer on Mercury would see the Sun slowly retrograde or "move backwards" through rather less than its own apparent diameter for eight Earth days around each perihelion passage. The Sun would then almost hover over what may be called a "hot pole". There are two hot poles, one or the other of which will always receive the full blast of solar radiation when Mercury reaches perihelion; the intensity will be $2\frac{1}{2}$ times greater than for regions of the surface 90 degrees away. Bear in mind, too, that from Earth the Sun has a mean apparent diameter of only about half a degree of arc; from Mercury the apparent diameter ranges between 1.1 and 1.6 degrees.

Let us consider two observers, both of whom are placed on Mercury's equator but who are 90 degrees away from each other in longitude. Observer A is at a hot pole, so that the Sun is at his zenith, or overhead

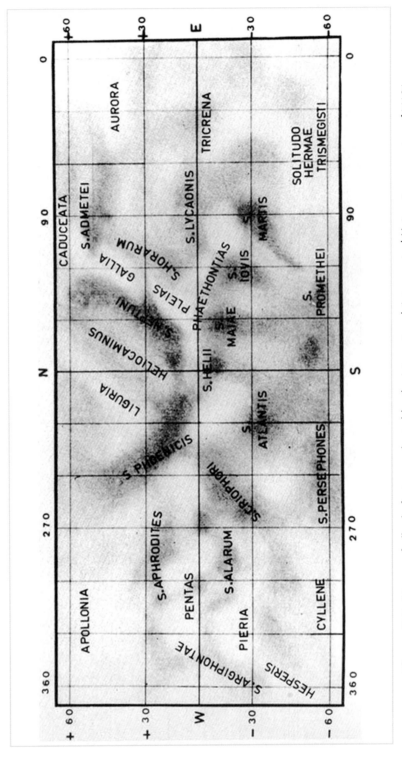

Figure 8.5. Names of albedo features adopted by the International Astronomical Union. Image courtesy of NASA.

point, at perihelion. This means that the Sun will rise near aphelion, and the solar disk will be at its smallest. As the Sun nears its zenith, it will slow down and grow in size. It will pass the zenith, and then stop and move backwards for eight Earth days before resuming its original direction of movement. As it drops toward the horizon it will shrink, finally setting 88 Earth days after having risen.

Observer B, 90 degrees away, will see the Sun at its largest near the time of rising, which is also Mercury's perihelion. Sunrise itself will be curiously erratic, because the Sun will come into view and then sink again until it has almost vanished. Then it will climb in the sky, shrinking as it nears the zenith. There will be no "hovering" as it passes overhead, but sunset will be protracted; Mercury is back at perihelion, so that the Sun will disappear, rise again briefly as though bidding adieu, and will then depart, not to rise again for another 88 Earth-days.

Note that because Mercury rotates three times for every two orbits, the same hemisphere will face the Sun at alternate perihelion passages. There are no seasons, because the tilt of the axis is effectively zero, and the polar regions remain cold. Another odd fact is that the stars will move across the sky at roughly three times the average rate of the Sun. One hates to think what a Mercurian would make of all this, but it is clear that no intelligent life can exist there, and in all probability no life of any kind.

That, really, was about as much as we knew before the Space Age. And so – on to 1973, and the epic journey of Mariner 10.

Figure 8.6. Scene on Mercury. Painting by L.F. Ball, with permission.

9 Mariner 10

Mariner 10 was a new kind of spacecraft. It was the first attempt to reach Mercury; there had been probes to Mars and Venus, and Pioneers 10 and 11 had already launched to the outer giants by the time that Mariner 10 soared aloft on 3 November 1973. However, the Atlas-Centaur launcher was not powerful enough to send Mariner direct to Mercury, and use had to be made of what is termed the gravity-assist technique. Venus would be encountered first, and its gravitational pull used to put Mariner into the correct course for Mercury.

Mariner was by no means huge. Its body was an eight-sided framework, measuring $4\frac{1}{2}$ feet diagonally and $1\frac{1}{2}$ feet in depth; the initial weight was 1108 pounds. Two solar panels, each 8.7 feet long and 3.2 feet wide, were attached at the top, supporting 55 square feet of solar cell area. Of course there was plenty of solar energy available in this part of the Solar System (with the space-craft sent to Jupiter and beyond, solar energy was in short supply, and the only answer was to use tiny nuclear power-plants). Mariner 10's launch was liquid-fuelled, and there was expected to be no trouble in keeping in constant radio touch with Earth. Navigating in space presents problems of its own, and Mariner carried a tracker which locked on to the brilliant southern star Canopus, second in brightness to Sirius.

I was among the spectators as Mariner was sent on its way from Cape Canaveral, and I was acutely conscious that I was watching history being made. All went well. After lift-off Mariner was put into a "parking orbit" round the Earth for just under half an hour, and then a burst of power from

Figure 9.1. Mariner 10. Image by courtesy of the Jet Propulsion Laboratory.

the motors sent Mariner into a path round the Sun. Its speed relative to the Earth had to be slowed down, which meant that the probe would spiral inward. All went well. The subsequent course of events had been very carefully planned:

8 February 1974. Encounter with Venus, passing the planet at a range of 2610 miles.

29 March 1974. First encounter with Mercury, passing over the surface at a distance of 438 miles before continuing in orbit round the Sun.

21 September 1974. Second encounter with Mercury, at a range of 29,200 miles.

16 March 1975. Third and last active pass of Mercury, at 203 miles above the surface. (It was unfortunate but unavoidable that at all three active passes the same areas would be sunlit, so that less than half the planet could be mapped. The rest would have to wait for a later mission.)

In the event the whole programme was carried through successfully. Contact with Mariner was finally lost on 24 March 1975, because the supply of gas needed to keep the instruments turned Earthwards was exhausted. Unless it has collided with some solid body, Mariner is still orbiting the Sun and making regular passes of Mercury, but we have no hope of contacting it again.

Various investigations were carried out during the journey to Mercury. Four thousand photographs were taken of Venus; it was confirmed that the rotation really is 243 days – longer than Venus' "year" of 224.7 days; the atmosphere is made up chiefly of carbon dioxide, with clouds rich in sulphuric acid; there is no detectable magnetic field. But of course the real excitement began when Mariner neared Mercury, and began to show details hopelessly beyond the range of any Earth-based telescope. It had been tacitly assumed that the surface would be of the same basic nature as that of the Moon, but nobody could be sure.

Figure 9.2. Venus, imaged from Mariner 10 on its way to Mercury. Image © Calvin J. Hamilton.

Mariner spent only seventeen hours close enough to Mercury to record really delicate detail, and all in all it was surprising that the results were as good as they actually were. The first surface feature was clearly seen six days before the first encounter; it took the form of a bright spot, later found to be a 60-mile crater with a system of rays not unlike those of the ray-centres on the Moon. The crater was later named in honour of Gerard Kuiper, the Dutch astronomer who had played a major rôle in planning the American programme of planetary research but who, sadly, had died just before the main results started to come through. Meanwhile, temperature measurements were being made, and it was also found that there is an appreciable magnetic field, certainly due to the presence of a large iron-rich core. There are no radiation belts comparable with the Van Allen belts which surround the Earth, but there is definite interaction between the Mercurian field and the solar wind, producing a well-defined bow shock.

The temperature measurements confirmed that in this respect Mercury is indeed a world of extremes. At a hot pole a thermometer would show a maximum value of 427 degrees Celsius, which is around 800 degrees

Figure 9.3. The bright, 60-mile crater named Kuiper (upper left). Image by courtesy of the Jet Propulsion Laboratory.

Fahrenheit; put a tin kettle on to the rocks, and the kettle will promptly melt. At night the temperature falls to –183 degrees Celsius, around –300 degrees Fahrenheit, so that the total range is well over 1000 degrees Fahrenheit, much greater than for any other planet in the Solar System. This is due to a combination of various factors: closeness to the Sun, the lack of a shielding atmosphere, and the slowness of the axial rotation, giving the surface plenty of time to heat up during the day and then an equal period of time to cool down at night. Also, the outer surface must be a "regolith" of relatively loose material similar to that found on the Moon, going down to a depth of a few feet.

However, there are places which remain bitterly cold, because they never receive any sunlight. And this leads me on to another point which is best discussed here, even though it may seem slightly out of context. It has been suggested that there may be ice on the floors of some of the polar craters of Mercury.

On the Moon we know of polar craters which are so deep that no sunlight can reach their floors, so that they remain numbingly cold – in fact, colder than anything we experience on Earth. In 1995 an unmanned lunar probe, Clementine (named after the character in the old mining song who was "lost and gone forever") sent back data indicating that the floors of some of these craters were ice-covered. Ice on the Moon sounded improbable, but the NASA announcement caused tremendous excitement, and it was claimed that there might be enough ice to provide sufficient water for a flourishing lunar-colony – even though the ice would certainly be difficult to extract from the rocks.

I was far from alone in being profoundly sceptical from the outset. The samples brought back from the Moon by the Apollo astronauts and the Russian unmanned probes had shown no trace of hydrated material; all were bone dry. And how could ice have found its way to the Moon? It could hardly be of lunar origin, and the only possible alternative seemed to be collisions with comets, but this is too hard to accept. An icy comet would smash into the Moon at high velocity, and would generate so much heat that the ice would be melted. Moreover, much of the resulting débris would be ejected from the Moon altogether (remember that the escape velocity there is a mere $1\frac{1}{2}$ miles per second). The idea of ice being conveniently deposited inside suitable craters seemed to be decidedly glib.

There was no claim that ice had been detected directly; what was said to have been found, by using an instrument known as a neutron spectrometer,[8]

----

[8] A neutron spectrometer is a powerful research tool. When cosmic rays from space hit the Moon, they produce neutrons, which can be detected. Collisions between cosmic ray particles and heavy atoms produce "fast" neutrons; if hydrogen atoms are hit, the neutrons are "slow". It was claimed that the neutron energy coming from the polar craters was of the slow variety, suggesting the presence of hydrogen and, hence, water ice.

was hydrogen. This could well be due to the solar wind, a stream of particles being sent out by the Sun in all directions all the time. The presence of ice had simply been inferred. At least the announcement did NASA's funds a great deal of good, and there were unkind hints that this might have been why some of the comments were so encouraging....

Three years later another space-craft, Prospector, went into orbit around the Moon and sent back reports which seemed to confirm those of Clementine. But doubts remained, and at last NASA decided to crash Prospector into one of the polar craters when its main programme had been completed – in the hope that water would be detected in the débris thrown up. Prospector did crash on schedule, but not a sign of water was found.

Efforts were made to confirm the existence of ice by using the giant radio telescope at Arecibo in Puerto Rico, much the most powerful in the world (its "dish" is 1000 feet across, and is built into a natural hollow in the ground). Initial results seemed to be promising, but it was then found that there were similar effects from regions which do receive sunlight and where ice could not possibly form. There the matter rests at present, but I doubt whether many people do now have much faith in lunar ice.

I have discussed this at some length because there have been similar claims about ice inside some of the polar craters of Mercury. However, the same objections apply, and it must be said that the presence of ice anywhere in Mercury is even less likely than it is for the Moon.

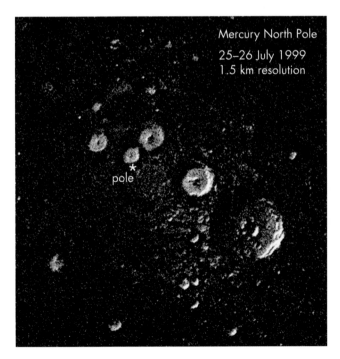

Mercury North Pole

25–26 July 1999
1.5 km resolution

pole

Figure 9.4. Ice? This image of Mercury's north pole was obtained by the Arecibo Radio Telescope in July 1999. The image is about 180 miles square, with resolution down to 1 mile. Many authorities believe that the bright pockets are due to water ice deposited by comets, though other people are decidedly sceptical! Image by courtesy of John Harmon, Arecibo Radar.

Figure 9.5. The Arecibo Radio Telescope. Photograph by R. Gill, with permission.

This has been something of a diversion, so let us now turn to another important investigation carried out by Mariner 10: the density and composition of the Mercurian atmosphere.

There were no real surprises. The ground pressure was found to be so low that the total weight of the whole atmosphere was only of the order of 8 tons. The main constituents were sodium, oxygen, helium, potassium and some argon; hydrogen and helium may well come from the solar wind, while sodium, potassium and oxygen are probably gases released by the vaporization of the surface layer (the regolith) by the impacts of tiny bodies called micrometeorites. As with the Moon, it is fair for most purposes to call Mercury an airless world.

All this was tremendously interesting, but the main task of Mariner 10 was to map the surface features. Within its limitations this part of the programme was outstandingly successful, and showed that Mercury was far from being the dull, uniform world that some people had expected.

10 Cratered World

It was unfortunate, but inevitable, that Mariner 10 imaged the same areas of Mercury at each of its three active passes. It therefore covered less than half of the total surface, and there is little hope of adding the rest from ground-based observations, plus the fact that the Hubble Space Telescope cannot be used (Mercury is always too dangerously near the Sun in the sky). It is particularly tantalizing that one of the main features, now called the Caloris Basin, was only half illuminated when Mariner passed over the planet. There is no reason to believe that the uncharted areas are basically different from those which we have seen, but of course we cannot be quite sure.

Before the Space Age we had the same sort of problem with the Moon, which does have captured or synchronous rotation; the same hemisphere is always turned toward the Earth, and 41 per cent of the surface is permanently out of view. Moreover, the areas near the edge of the Earth-turned side are highly foreshortened, so that it is difficult to tell a crater from a mountain or a ridge. My own observing effort was concentrated upon mapping these foreshortened areas, carried in and out of view by "librations", so that I came to know them very well. All sorts of theories had been advanced about the far side. One nineteenth-century astronomer even suggested that all the Moon's air and water had been drawn round there, so that there might after all be advanced life. More rationally, it was also suggested that the arrangement of the features might be different from the areas we know, and it was pointed out that with one exception all the

Figure 10.1. The lunar Mare Orientale, probably the nearest lunar equivalent to Mercury's Caloris Basin. Image courtesy of NASA.

important seas or maria kept to the near side, and did not extend over the limb.[9] When the first images of the far side were obtained, by the Russian space-craft Lunik 3 in 1959, it became clear that although the newly-seen regions were just as crater-scarred, mountainous and barren as those we had always known, there were distinct differences; for example, there were no major maria wholly on the far hemisphere, there were numbers of "palimpsests", large enclosures with light floors and very low walls. We may be in for some major surprises when we examine the uncharted part of

---

[9] The exception was the Mare Orientale or Eastern Sea, a very small part of which can be seen under the best possible conditions of libration. I am proud to say that I discovered it myself, in 1948, using my modest $12\frac{1}{2}$-inch refractor, and I even proposed its name, though I thought that it must be a small mare; I had no means then of telling that it is really a vast, complex impact structure. Some years later it was rediscovered by American observers, but my suggested name was retained.

Mercury, but somehow I rather doubt it. At least we ought to find out when the latest probe, Messenger, arrives there in 2011.

We have already had one surprise. Though in many ways the surface of Mercury does resemble that of the Moon, there are very marked differences. Mercury has no lunar-type maria, and the Moon has no Mercury-type "intercrater plains" of any size, but craters dominate both surfaces, and when one crater breaks into another it is almost always the smaller crater which is the intruder. Cases of small craters being invaded by larger ones are very rare indeed, and this is understandable, since it is fairly evident that the large craters must be older than the smaller formations.

The features shown on Mercury by Mariner 10 fall into several well-defined categories. Craters less than about 13 miles across are usually bowl shaped, with depths of about 1/5 of their diameters; craters between 13 and 56 miles across had flatter floors, often with central peaks and terraced walls. Some craters are larger than any to be found on the Moon, and one, Beethoven, is a full 400 miles in diameter.

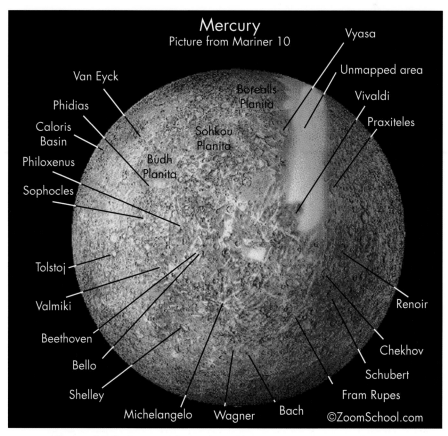

Figure 10.2. Principal features on Mercury. Image courtesy of NASA.

Naming the newly-seen features was the responsibility of the Nomen-clature Commission of the International Astronomical Union, the control-ling body of world astronomy. (I was for many years a member of that Commission, though in 2001 I had to retire because I was no longer well enough to attend meetings overseas.) As well as craters, there were plains (Latin, *planitiæ*); ridges or scarps (*dorsa*), valleys (*valles*) and mountains (*montes*). As with the Moon, craters were named after famous people, though not necessarily scientists; there are musicians (Beethoven, Sibelius, Chopin), authors (Shakespeare, Homer, Ibsen) and so on. There are few astronomers, apart from Kuiper, whose crater was actually the first to be recognized as Mariner 10 drew in toward its target; we all regretted that Gerard Kuiper, who had played so major a rôle in all the planning, was not able to be with us. Plains are named after the word for 'Mercury' in differ-ent languages, such as Odin Planitia (Scandinavian) and Suisei Planitia (Japanese), though there are two exceptions: Borealis Planitia (the Northern

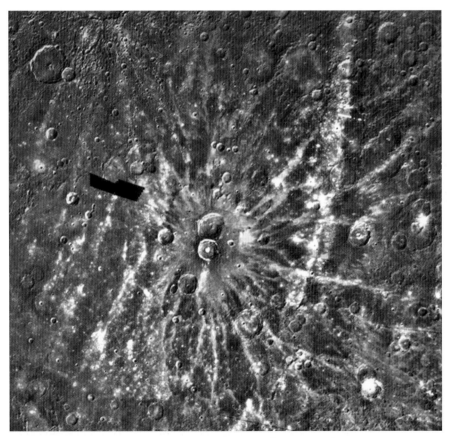

Figure 10.3. The crater Degas, in Mercury's north-west quadrant. Image courtesy of NASA.

Plain) and Caloris Planitia (the Plain of Heat). Caloris was so named because it contains one of the "hot poles" – a sort of Mercurian Death Valley. Scarps are generally named after ships which have been involved in exploration or scientific research, such as *Discovery* and *Santa Maria*, though two are named after astronomers, Schiaparelli Dorsum and Antoniadi Dorsum. Valleys are named after important radio telescopes (Arecibo Vallis, Goldstone Vallis). On the whole the system works well. Wisely no post-1700 politicians or military leaders are allowed on Mercury. It would undeniably be awkward to have adjacent craters named after Winston Churchill and Adolf Hitler!

The next step was to draw up a system of co-ordinates, and this too was the responsibility of the International Astronomical Union. One early suggestion was to make the prime meridian by the longitude of the sub-solar point at the time of the first perihelion passage of Mercury after 1 January 1950, but this was a problem because that longitude was not imaged by Mariner: at each active pass it was in darkness. However, a small crater now called Hun Kal, almost a mile in diameter was reckoned to be on the 20-degree meridian as defined by the I.A.U., and this was used as the reference. (Why Hun Kal? Because this was the Mayan word for "twenty":

Figure 10.4. Crater Chao Meng-fu (in shadow, upper centre), which contains Mercury's geographical south pole. Image courtesy of NASA.

the Maya were the most erudite of all the astronomers of pre-Spanish America, and their system of numbers was based upon 20 rather than 10.) The geographical[10] south pole lies inside a larger crater, Chao Meng-fu, which was available to Mariner.

There is no doubt that the craters are impact structures, and there are "secondary" craters, due to the impacts of material hurled out from the first strike. These secondaries, and also other ejecta deposits, are always found near the relevant main crater, but they are closer-in than those on the Moon, because Mercury has a greater force of gravity, and material flung out after the primary impact cannot fly so far. Much the largest formation is the Caloris Basin (Caloris Planitia), over 800 miles across, and flanked by mountains and "hummocky" terrain; it is a great pity that half of it was in darkness when Mariner passed over. It is filled with material making up a relatively smooth surface or plain, and in some respects it seems to be not too unlike the Mare Orientale on the Moon.

There are some notable ray-systems, such as those centred on the craters Kuiper and Snorri. Presumably they are of the same nature as the ray-systems on the Moon, but there are none anything like so large or so striking as those of Copernicus and Tycho on the lunar surface. However, they cut across and are superimposed upon all other surface features, so that the ray-craters themselves must be the youngest craters anywhere on Mercury.

The broad zones between the heavily-cratered areas are known as intercrater plains. They lack major structures, but have many small craters below 6 miles in diameter, some of which are shallow and elongated, and may well be secondaries. In general the intercrater plains show gentle undulating relief, "rolling ground" between the heavily cratered regions. Hummocky plains lie within a ring 350 to 500 miles wide, circumscribing the Caloris Basin; they contain low, scattered hills, and seem to be due to material ejected after the Caloris impact. There are also smooth plains, relatively level and lightly cratered; they are found outside Caloris as well as in some of the smaller basins, notably Borealis Planitia in the far north. They seem to be the closest analogies to the lunar maria, and we may well assume that the Caloris event triggered off violent vulcanism which lasted for some time.

It is also notable that on the exactly opposite region of Mercury – the "antipodes" of Caloris – we find what is known officially as hilly and lineated terrain, though it is known to most people as "weird terrain", multi-grooved, with smooth strips between the grooves. There seems little doubt that this was caused by shock-waves resulting from the Caloris impact; in

---

[10] This really should be "hermographical", from the Greek *Hermes*, but I doubt whether this term is ever used nowadays.

Figure 10.5. Mariner 10 mosaic of Mercury taken during the approach to the first encounter, made up of 18 images taken at 42-second intervals 6 hours before closest approach. The lower two-thirds of the visible portion is Mercury's southern hemisphere. Resolution is about $1\frac{1}{4}$ miles (2 km). The flat-floored crater just below centre, containing two smaller craters at its bottom right and left, is the 100-mile (160 km) diameter Petrarch crater. The light-floored crater surrounded by darker material at the upper left is Lermontov, with a diameter of 100 miles (160 km). Image courtesy of NASA.

fact a sort of "lensing" effect. Indications of similar effects are found on the Moon, antipodal to Mare Imbrium and Mare Orientale.

The Moon lacks large intercrater plains (which are more cratered than the smooth plains, and must therefore be older). Also, there are no lunar features comparable with the "lobate scarps" on Mercury. These are lobed, rounded fronts winding across the surface. In some cases they extend for hundreds of miles, often up to 40 miles broad and at least a mile high. They too must be young by Mercurian standards, because they often cut through

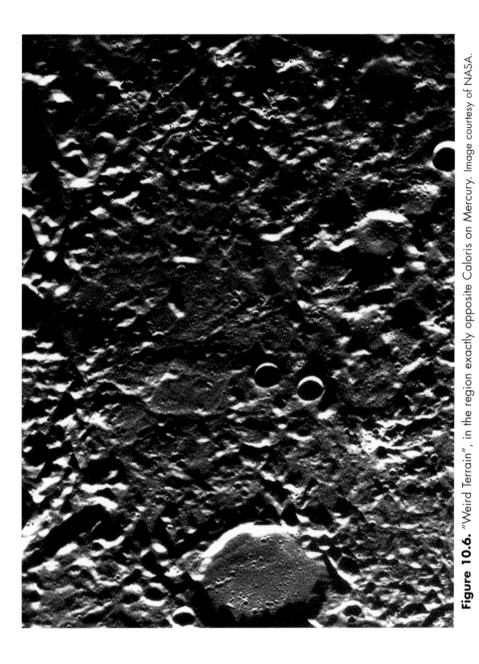

Figure 10.6. "Weird Terrain", in the region exactly opposite Caloris on Mercury. Image courtesy of NASA.

craters. They seem to have been formed as Mercury cooled and the crust shrank, wrinkling into the curving scarps. The largest lobate scarp found on Mercury is Discovery Rupes, 350 miles long and one mile high; the walls and floors of craters have been deformed by the thrust fault which formed the scarp. At this stage it may be useful to say a little about faulting in general, because it is so vitally important to any geologist.

There are three main types. Normal faulting occurs when the crust is being stretched; if the stress becomes greater than the strength of the rocks, a block of material will drop down. If the crust is being compressed, one block of material will be pushed over another (thrust fault). With a strike-slip or transverse fault, the two blocks of material simply slide past each other. Transverse faults are very common on the Earth, but much less so on Mercury and the Moon; the Mercurian lobate scarps are due to thrust-faulting. Elsewhere a section of crust will slide down between the two oppositely-facing normal faults, producing what is known as a graben.

Obviously we know less about the past history of Mercury than we do about that of the Moon, but at least we have some facts to guide us. It is safe to assume that Mercury and the Earth formed at the same time (4.6 thousand million years ago) and began to evolve along the same lines, though there were important differences because Mercury is so much the smaller and less massive of the two globes. Both were formed from the hot, inner regions of the solar nebula. There is little doubt that Mercury once rotated much more quickly than it does now; it was slowed down by tidal forces, until relative to the Sun (not relative to the stars) the rotation became equal to two-thirds of a Mercurian year. Initially Mercury was hot and viscous, and quite apart from the accretion process we may also have to consider the heat produced by radioactive elements; much of the globe may have been melted. The heavier elements such as iron sank downward, and, as we have noted, the heavy iron-rich core of today is probably larger than the entire body of the Moon; it was here that the magnetic field developed. The crust was ripped apart and lava poured out through the fractures. This was the time of the "great bombardment"; the scars remain on both Mercury and the Moon, though not on Earth, and Venus' dense atmosphere makes it a special case.

As Mercury cooled down, the crust thickened and volcanic action became less violent. There was global contraction, producing the lobate scarps, but the Caloris impact – probably four thousand million years ago – had major effects everywhere, producing the "weird" hilly and lineated terrain at its antipodes. The rotation continued to slow down, and in the end Mercury became the placid, inactive world that we know.

Two extra points are worth noting. First, there is no doubt that the history of the Earth has been strongly moulded by the presence of the

Moon; whether the giant impact theory is correct, or whether the Earth and the Moon were both formed in the same part of the solar nebula, is irrelevant in this context. It is the Moon which has stabilized the Earth's axis of rotation and kept it close to the present day value of $23\frac{1}{2}$ degrees to the perpendicular relative to the orbital plane. Otherwise there would have been marked changes in the angle of the tilt, causing most uncomfortable variations in climate. Mercury has no satellite. There was a flurry of excitement on 27 March 1974, two days before the fly-by of Mariner 10, because one of the instruments began to record bright emissions at ultra-violet wavelengths. They vanished, but then reappeared, and the presence of a satellite was suspected. Disappointingly, it turned out that the culprit was an ordinary star, 37 Crateris. If Mercury did have a satellite of appreciable size, it would certainly have been found by now.

Secondly, we have to explain the disproportionately large size of the core. It has been suggested that at one stage Mercury was struck by a large asteroid, so that its outer shell was "peeled off" and what we see now is simply the core of the original planet. This may or may not be correct, but in any case Mercury's overall density is much the same as that of the Earth, and much higher than for any other planet or satellite in the Solar System.

We can attempt to draw up a table for the history of Mercury, and five epochs are accepted, each named after a prominent formation – Caloris, and three craters, Tolstoj,[11] Mansur and Kuiper:

Epoch	Age, thousands of millions of years ago	Features	Lunar counterpart
Pre-Tolstojan	Over 4	Intercrater plains, multi-ring basins	Pre-Nectarian
Tolstojan	4	Plains materials, smaller basins, craters	Nectarian
Calorian	4–3	Basins, plains materials	Imbrian
Mansurian	3–2	Craters	Eratosthenian
Kuiperian	1	Craters, ray-systems	Copernican

There seems to be no chance of active vulcanism on Mercury now. There may be occasional impacts, bearing in mind that comets are frequent visitors to these regions of the Solar System, and there may be numbers of Vulcanoid asteroids, but whether we could identify a fresh crater seems to

---

[11] This is, of course, Tolstoy. Why he becomes Tolstoj on Mercury I know not.

Figure 10.7. Comet C/1999 S4 (Linear) on July 22, 2000. Image courtesy of John Fletcher.

be doubtful, though we cannot rule it out if we undertake long-term surveys with orbiting satellites. But to all intents and purposes, Mercury has come to the end of its active career.

11 Around Mercury

A general look around Mercury is interesting, if rather restricted. For the Moon and Mars, the user of a small telescope can go out and see many of the surface features for himself; for Mercury this is obviously impossible, and since I doubt whether any reader can afford to hire a space-craft I have made no attempt here to give anything in the nature of a conventional atlas, particularly as we know nothing as yet about the areas not imaged by Mariner 10. In general I have listed only the main features, with craters over about 90 miles in diameter plus a few more of special interest (such as ray-centres).

In the official IAU map of Mercury, the accessible surface is divided into "quadrangles", not all of which are fully covered and not all of which are quadrangular. The names are linked with the old Antoniadi system, which is very approximate but better than nothing at all. The NASA quadrangles are named after a particular feature inside. (See table at the top of page 88.)

Two other quadrangles, H-4 (Antoniadi's Liguria) and H-13 (Solitudo Persephones) had only about 10 degrees covered.

For our present purpose it may be better merely to give a brief description of the imaged area, divided into four quadrangles.

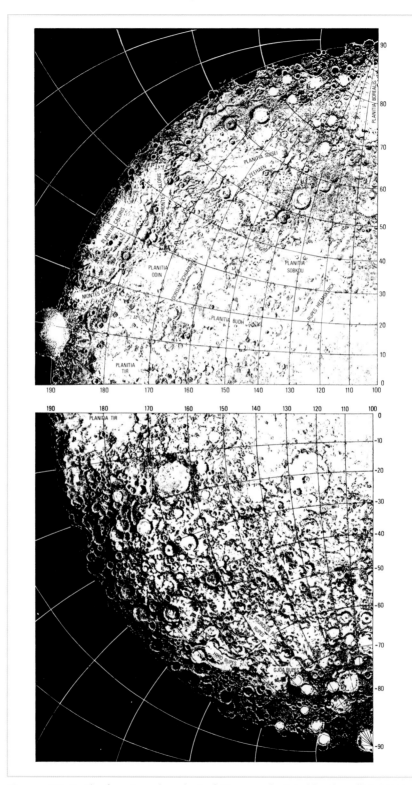

Figure 11.1. The four "quadrangles" of Mercury depicted by the official IAU map.
Courtesy of the National Space Science Data Center.

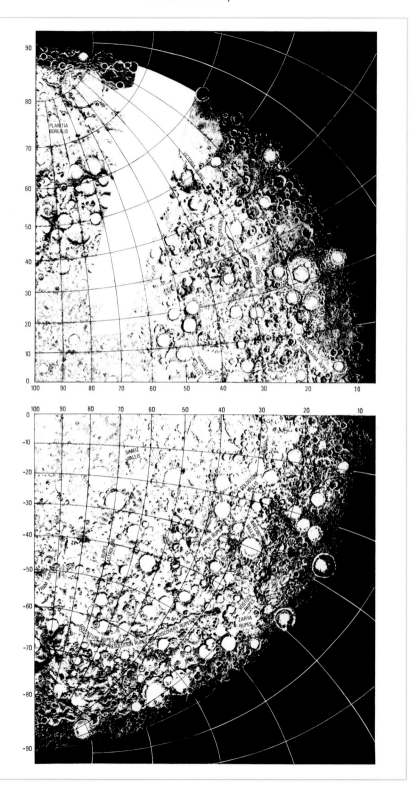

Quadrangle Name	Selected Feature	Antoniadi Equivalent
Borealis	Borealis Planitia	Borea
Victoria	Victoria Rupes	Aurora
Shakespeare	Shakespeare (crater)	Caduceaya
Kuiper	Kuiper (ray-crater)	Tricrena
Beethoven	Beethoven (crater)	Solitudo Lycaonis
Tolstoj	Tolstoj (crater)	Phæthontias
Discovery	Discovery Rupes	Solitudo Hermæ Trismegisti
Michelangelo	Michelangelo (crater)	Solitudo Promethei
Bach	Bach (crater)	Australia

North-West Quadrant

	N. Lat.	DORSUM Long.	Name
Schiaparelli	24	164	G. V.: Italian astronomer (1870–1944)
		MONTES	
Caloris	22–40	180	(Heat)
		PLANITIA	
Budh	18	150	Hindu for "Mercury"
Caloris	30	199	(Heat)
Odin	24	171	Norse god
Sobkou	39	128	Messenger god
Suisei	58	157	Japanese messenger god
Tir	03	177	Norse for "Mercury"
		RUPES	
Heemskerck	27	125	Dutch; one of Tasman's ships
Zeehaen	50	158	Dutch; one of Tasman's ships

	N. Lat.	Long.	CRATERS Diameter, miles	Name
Durer	22	119.5	118	Albrecht; German painter (1471–1528)
Lysippus	1.5	133	93	Greek sculptor (4th cent. BC)
Mozart	8	191	140	Austrian composer (1756–1791)
Phidias	9	150	96	Greek sculptor (c. 490–430 BC)
Verdi	65	169	93	Giuseppe: Italian composer (1813–1901)
Wang Meng	10	104	108	Chinese painter (1308–1385)
Brahms	59	177	75	Johannes; German composer (1833–1897).

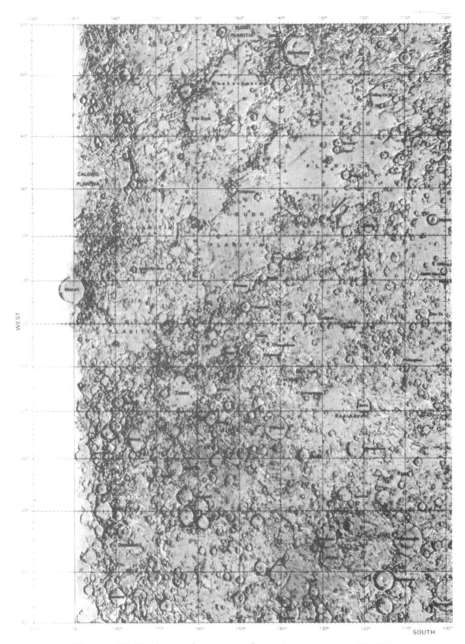

Figure 11.2. The north west quadrant. Image courtesy of NASA.

The whole topography of this quadrant is dominated by the Caloris Basin. Only part of it is shown; the rest was unavailable to Mariner 10. It is comparable in size with the lunar Mare Imbrium, and the resemblance may be more than superficial. The effects of the Caloris impact extend far

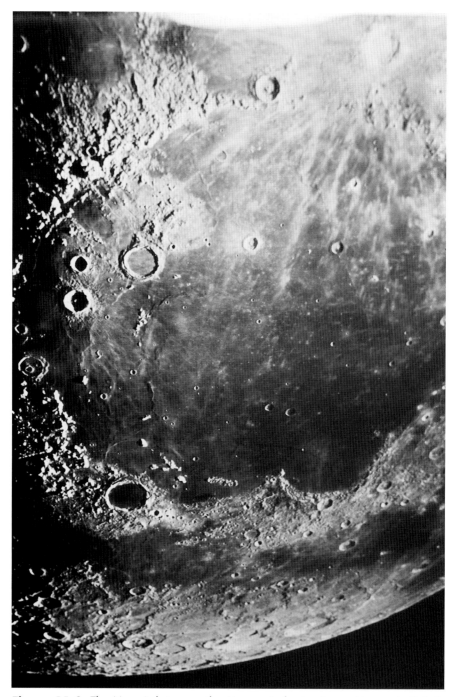

Figure 11.3. The Mare Imbrium on the Moon, similar in size to Mercury's Caloris Basin. Photograph by Patrick Moore, using his 15-inch reflector.

beyond this quadrant, just as the Imbrium impact affected wide areas of the lunar surface.

Outside the great basin we have areas flooded with smooth plains material; others are covered with rougher material, making up the so-called hummocky terrain. This area extends for over 600 miles around Caloris, and contains five of the named plains: Tir, Budh, Odin, Sebkou and Suisei. The effects of the Caloris impact decrease with increasing distance, finally merging into inter-crater type of terrain so common over much of Mercury.

In some regions it seems that older features have been completely obliterated, but here and there, round the margins of the smooth plains material, lunar-type ghost craters can be made out.

Figure 11.4. The Caloris Basin (at left). Image courtesy of NASA.

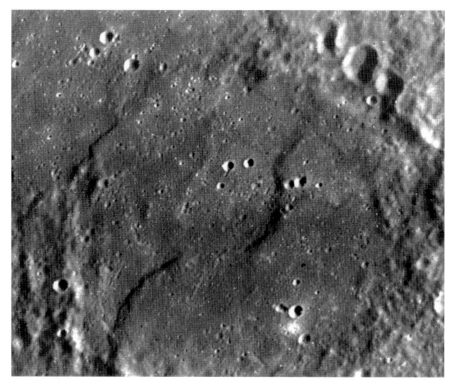

Figure 11.5. Mariner 10 image showing scarp features in the south-east Tir Planitia region of Mercury. These scarps may be volcanic flow features or compressional faults. This image was taken during the first fly-by; north is up, and the scarps trend from south-west to north-east. The frame is about 130 km across. Image courtesy of NASA.

Mozart is an interesting crater; it was clearly formed after the Caloris Basin, and is probably one of the youngest craters on the mapped part of the surface. Its west wall is the illuminated area which reached furthest into the night hemisphere.

There are several ray-craters in this quadrant, and two named mountain ranges, Heemskerck and Zeehaen.

North-East Quadrant

The coverage here is incomplete, because one strip of the surface was inaccessible to Mariner 10. There is no reason to believe that the missing area contains any special features – though, of course, one never knows!

Figure 11.6. Crater Brahms (at upper right). Image courtesy of NASA.

	N. Lat.	DORSUM Long.	Name
Antoniadi	28	30	Greek astronomer (1870–1944)
		PLANITIA	
Borealis	75	85	North
		RUPES	
Endeavour	39	31	Cook's ship to go to Tahiti
Santa Maria	6	20	Columbus' flagship, 1492
Victoria	50	32	Spanish; Magellan's ship
		VALLIS	
Haystack	4	46	Radio telescope, Massachusetts, USA

	N. Lat.	Long.	CRATERS Diameter, miles	Name
Al-Hamadhani	39	89	105	Arab writer (?–1007)
Chaikovskij	8	51	100	P. Tchaikovsky, Russian composer (1840–1893)
Derzhavin	45	36	90	G.R.: Russian poet (1743–1816)
Giotto	13	56	93	Italian painter (c. 1271–1337)
Goethe	80	44	211	J. von: German poet (1749–1832)
Handel	4	34	93	G.: German composer (1685–1759)
Homer	0	37	199	Greek poet (9th century BC)
Hugo	39	48	118	V.; French writer (1802–1885)
Kuan Han-ch'ing	79	53	96	Chinese dramatist (1241–1320)
Lermontov	16	50	99	M.Y.; Russian poet (1814–1841)
Melville	22	10	84	Herman; American author (1819–1891)
Monet	44	10	155	Claude; French painter (1840–1926)
Praxitiles	27	60	102	Greek sculptor (c. 370–330 BC)
Rodin	22	19	168	Auguste: French sculptor (1840–1917)
Rubens	60	74	112	Peter: Flemish painter (1577–1640)
Sholen Aleichem	11	77	118	Yakov Rabinowitz: Yiddish writer (1859–1916)
Stravinsky	50	73	106	Igor: Russian composer (1882–1971)
Vivaldi	15	86	130	Antonio: Italian composer (1678–1741)
Vyāsā	49	80	171	Indian poet (c. 1500 BC)
Wren	24	35	133	Christopher: English astronomer/architect (1632–1723)

This quadrant contains some smooth plains material, plus some ray-craters and the Antoniadi Dorsum, which seems to be a southern extension of the Endeavour Rupes; the scarp has turned into a ridge simply because the level of the upthrust side of the fault has developed more abruptly. The large crater Goethe may be geologically related to the Borealis Planitia.

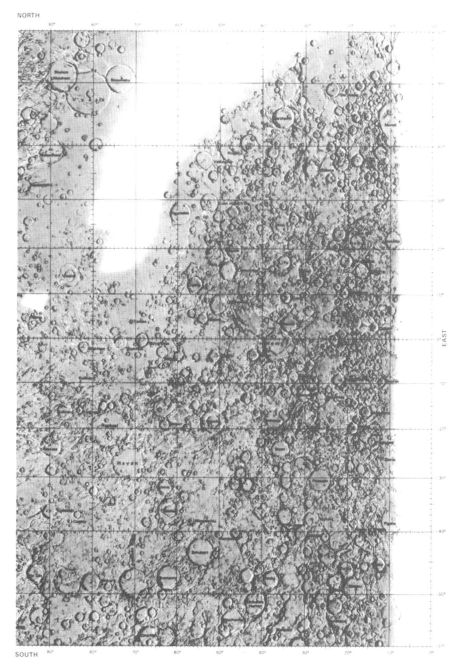

Figure 11.7. The north-east quadrant. Image courtesy of NASA.

Rodin is a large and well-developed double-ring basin, associated with a wide area of ejecta and secondary cratering. Right on the terminator is another very well-formed crater, Melville. Of the four named valleys, one (Haystack) is really a crater-chain.

Figure 11.8. The Antoniadi Dorsum. Image courtesy of NASA.

South-East Quadrant

	S. Lat.	Long.	Name
			RUPES
Adventure	64	63	English: one of Cook's ships, 1772–1775
Astrolabe	42	71	French; D'Urville's ship to explore Antarctica 1838–40
Discovery	54	38	English; one of Cook's ships, 1776–80
Fram	572	94	Norwegian; Nansen 1892–96, Sverdrup and Nansen 1909
Mirni	37	40	Russian; Bellinghausen's ship, 1819–21
Resolution	62	52	English; one of Cook's ships, 1772–75
Vostock	38	19	Russian; Bellinghausen's ship, 1819–21
Zarya	41	22	USSR; schooner to study Earth's magnetic field, 1953
			VALLIS
Arecibo	27	29	Radio telescope in Puerto Rico
Goldstone	15	32	Radio telescope in California, USA
Simeiz	13	68	Radio telescope in Crimea, Ukraine

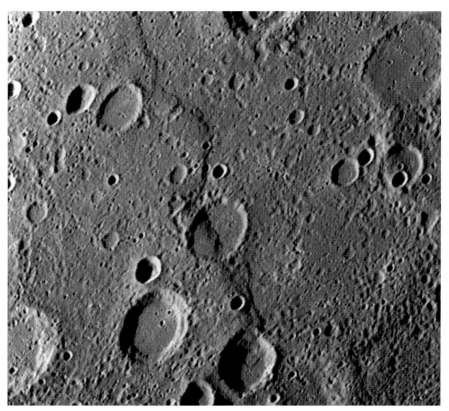

Figure 11.9. Santa Maria Rupes, taken by Mariner 10. Image courtesy of NASA.

	S. Lat.	Long.	CRATERS Diameter, miles	Name
Brunelleschi	9	22	87	Filippo; Florentine architect (1377–1446)
Chekhov	36	62	112	Anton Tchekhov, Russian writer (1860–1904)
Copley	39	86	19	John: American painter (1738–1815)
Haydn	28	70	143	J.; Austrian composer (1732–1809)
Homer	0	39	199	Greek poet; 8th or 9th century B.C.
Ibsen	24	36	100	Heinrich: Norwegian dramatist (1828–1906)
Imhotep	18	39	100	Egyptian physician (c. 2686–2613 B.C.)

(Continued on page 98)

	S. Lat.	Long.	CRATERS Diameter, miles.	Name
Kuiper	11	32	25	Dutch astronomer (1905–1973)
Kurosawa	52	23	112	K.; 18th-century
Ma Chih-Yuan	61	77	106	Chinese dramatist (c. 1251)
Matisse	24	90	130	Henri; French painter (1869–1954)
Mendes Pinto	61	19	106	Portuguese author (c. 1510–1583)
Petrarch	30	27	99	Francesco; Italian poet (1304–1374)
Pigalle	38	12	81	Jean; French sculptor (1714–1785)
Pushkin	66	23	124	Alexander; Russian poet (1799–1837)
Raphael	21	77	217	Raffaello Sanzio; Italian painter (1483–1520)
Renoir	18	52	137	Pierre; French painter (1841–1919)
Schubert	42	55	99	Franz; Austrian composer (1797–1828)
Sotatsu	48	19	81	Tawaraya; Japanese artist (1600–1643)
Snorri	8.5	84	12	Sturluson; Icelandic poet and writer (1179–1241)

Most of this quadrant is heavily cratered; there are also some areas of intercrater plains pitted with small secondary craters. The floors of many of the large craters have been flooded with small plains material; good examples of this are Pushkin and its companions Mendes Pinto, Raphael, Petrarch and Haydn. There is also Pigalle, which is too near the terminator to be clearly shown.

The quadrant includes the ray-crater Kuiper, the first feature to be identified during the Mariner 10 approach, and two other ray-craters, Snorri and Copley. The scarps include Discovery Rupes, the most prominent feature of its kind on the accessible region of Mercury. There are also three valleys, which are not alike: Goldstone and Arecibo are chains of secondary craters, not too unlike lunar features such as the Rheita Valley, but Simeiz Vallis may be simply a prominent terrace on the wall of a large unnamed crater.

The area between latitudes 20 and 40 south and longitudes 10 to 40 degrees west is characterised by "hilly and lineated" terrain, often called

Figure 11.10. The Vostock Rupes region, with Petrarch the flat crater just left of centre. Image courtesy of NASA.

"weird terrain". It is antipodal to the Caloris Basin, and is thought to be due to seismic waves focused through the globe of Mercury at the time of impact. It has been estimated that the impactor was perhaps 30 miles across, so that if anything of comparable size landed on the Earth it would cause a tremendous amount of damage.

Much of this quadrant is occupied by intercrater plains: there is one region of smooth plains – actually the southern extension of Tir Planitia, most of which lies in the north-east quadrant. There are two huge craters with floors flooded with smooth plains material, Beethoven and Tolstoj; Beethoven is the largest formation of its kind on Mercury, unless one includes the Caloris Basin. Beethoven contains several craters, including Bello.

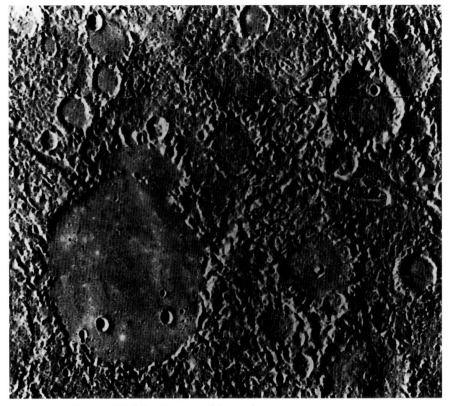

Figure 11.11. Close-up of Petrarch. Image courtesy of NASA.

South-West Quadrant

	S. Lat.	Long.	RUPES Name
Gjoa	65	163	Norwegian; Amundsen's ship, 1903–1906.
Hero	57	173	American; Palmer's ship, 1820–1821.
Pourquoi-Pas	58	156	French; Charcot's ship, 1908–1910.

Figure 11.12. The ray-crater, Copley (at upper left). Image courtesy of NASA.

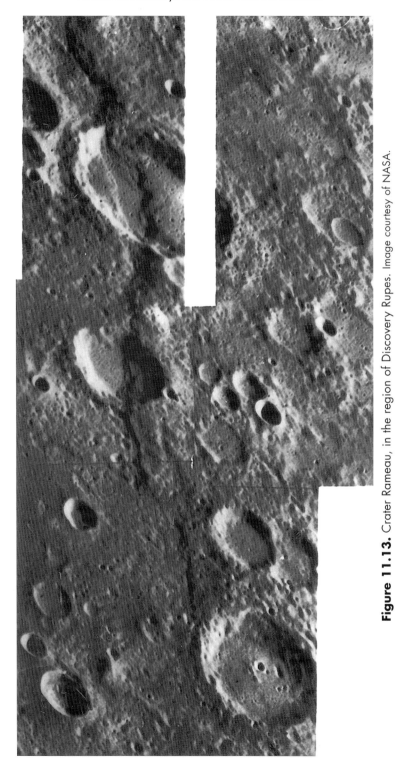

Figure 11.13. Crater Rameau, in the region of Discovery Rupes. Image courtesy of NASA.

	S. Lat.	Long.	CRATERS Diameter, miles	Name
Bach	69	103	140	J.S.; German composer (1685–1750)
Beethoven	20	124	388	Ludwig van; German composer (1770–1827)
Bello	19	121	93	Andres; Venezuelan poet (1781–1865)
Cervantes	76	122	124	Miguel; Spanish writer (1547–1616)
Chao Meng-fu	88	132	93	Chinese painter (1254–1322)
Lysippus	2	133	93	Greek sculptor (4th century B.C.)
Mark Twain	11	140	87	Samuel; American novelist (1835–1910)
Milton	26	175	109	John; English poet (1608–1674)
Sophocles	7	147	90	Greek dramatist (c. 496–406 B.C.)
Tolstoj	15	165	248	Lev Tolstoy; Russian novelist (1828–1910)
Vālmik	20	142	137	Sanskrit poet (1st century B.C.)
Wagner	68	114	84	Richard; German composer (1813–1881)

The south polar area is very rough, and contains several prominent craters, including Cervantes. The south polar point lies just inside the rim of the crater Chao Meng-fu, the floor of which is in permanent shadow.

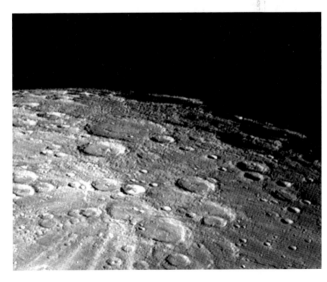

Figure 11.14. Crater Chao Meng-fu (in shadow, upper centre). Image courtesy of NASA.

This has been a very sketchy account, but it may serve to show the nature of the terrain imaged by Mariner 10. When our next probe reaches the planet we will have a great deal of extra information, and certainly our maps will have to be drastically revised, but at least a start has been made.

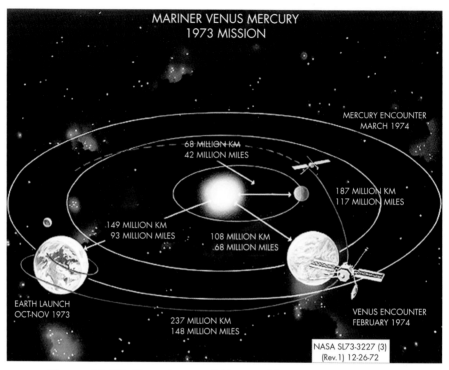

Figure 11.15. The trajectory of Mariner 10. Image courtesy of NASA.

12 Return to Mercury

For some time during the last decades of the twentieth century and the first years of the twenty-first, Mercury was often called "the forgotten planet". With the main emphasis on Mars, and forays also to Jupiter and Saturn, Mercury was indeed neglected, no doubt because the chances of finding life there were effectively nil. Neither was Venus in the forefront of space planning; moreover funds for scientific research were being regularly cut. Yet eventually there was a slight change of heart, and both Venus and Mercury were again scheduled to be visited by space-craft.

Perhaps surprisingly, Mercury took precedence. The Venus Express was not launched until 2005, but the Messenger to Mercury was seen on its way on 3 August 2004. It was launched from Cape Canaveral on a Boeing Delta-2 rocket. There was one initial disappointment. Originally it had been planned to send not only an orbiter, but also a lander to come down gently on the planet and send back data direct from the surface. This seemed to be practicable, and after all the Mars landers, Spirit and Opportunity, were amazingly successful. It is true that the Martian atmosphere was a definite help, but from a technological point of view there seemed to be no reason why landings on Mercury would present very severe problems. Sadly, the accountants stepped in yet again. Messenger survived, but it lost its lander, and was reduced to a straightforward orbiter. The scientists accepted this as gracefully as they could, and determined to make the best possible use of the watered-down mission.

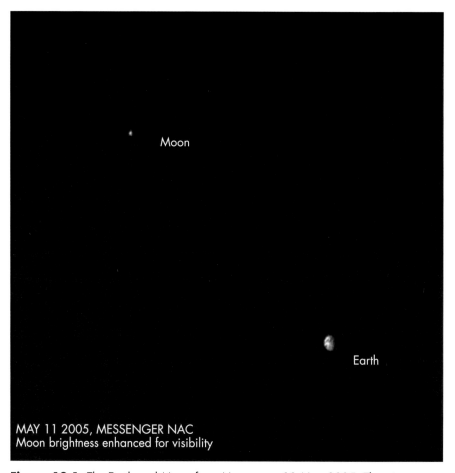

MAY 11 2005, MESSENGER NAC
Moon brightness enhanced for visibility

Figure 12.1. The Earth and Moon from Messenger, 11 May 2005. The picture was obtained with MDIS (Messenger Dual Imaging System), range 18,400,000 miles. Image courtesy of NASA/Johns Hopkins University Applied Physics Laboratory/Carnegie Institution of Washington.

Despite having imaged less than half the planet, Mariner 10 had been a great success, and some people questioned whether a future unmanned probe could tell us much more. Yet there was a great deal that we did not know, and in addition to mapping the rest of the surface Messenger had some very definite goals. We wanted to know:

1. Why is Mercury so dense?
2. What is its geologic history?
3. What is the structure of Mercury's core? Is it liquid, and if so how large is the liquid component?
4. What is the nature of the magnetic field – is it in any way like that of the Earth, or does it arise from a completely different process?

5. Is there ice inside the polar craters?

6. How has the thin atmosphere evolved? Six gases were known – hydrogen, helium, oxygen, sodium, potassium and calcium – but the density is so low that astronomers usually avoid the term "atmosphere", preferring "exosphere".

There was a good deal of disagreement. For instance, it was quite widely believed that ice did exist on the floors of polar craters such as Chao Meng-fu, where sunlight never penetrates, but there were total sceptics – myself, for example, and (much more significantly) Harrison Schmitt, who has the advantage of having been to the Moon. Also, if Mercury has a liquid core and magnetic field, why is the surface so inert? And how important has volcanic action been throughout the planet's history? Messenger would, it was hoped, provide at least some of the answers.

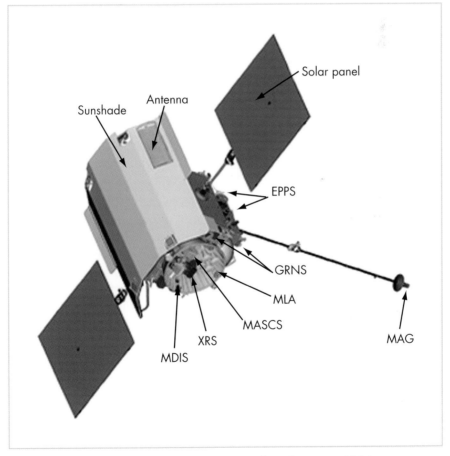

Figure 12.2. Messenger. (Abbreviations are explained on page 108.) Image courtesy of NASA.

There was nothing really revolutionary about the design of Messenger, apart from the addition of a sunshade to guard against the intense heat. The motors were of the conventional liquid-fuel variety, not the more advanced propulsion system used to send Europe's SMART-1 vehicle to the Moon. (SMART had taken a year to reach lunar orbit, and was just about arriving there when Messenger soared aloft.)

It was said that Messenger's body was in the form of a squat box, measuring $56 \times 73 \times 50$ inches, which is not very much. The sunshade measured 8×6 feet, and the solar panels extended for 20 feet from end to end; the "dry weight" was 1100 pounds, while the fuel accounted for another 1300 pounds. The total cost was 427 million dollars. This sounds a great deal – until it is compared with the cost of, say, a nuclear submarine.

The instruments were chosen so as to obtain as much varied information as possible (see Figure 12.2). The Mercury Dual Imaging System (MDIS) consists of two cameras which will image landforms, track variations in surface spectra and gather topographic information; a stunning mirror will help to point the instrument in any chosen direction. The Gamma-Ray and Neutron Spectrometer (GRNS) will detect gamma-rays and neutrons which are emitted by radioactive elements on Mercury's surface, or by surface elements which have been stimulated by cosmic rays (which are bombarding Mercury all the time). This will quite possibly clear up the question of polar ice. The X-ray spectrometer will identify elements on the surface. The Magnetometer (MAG) is sited at the end of a 12-foot boom, and will search for magnetized rocks in the crust as well as studying the global magnetic field. Mercury Laser Altimeter (MLA) contains a laser which will send light to the planet's surface, and a sensor which will receive the "echo" when the light is reflected back; together they will measure the time taken for a "round trip" to the surface and back, and recording variations in this distance will produce very accurate measurements of Mercury's topography. The Energetic Particle and Plasma Spectrometer (EPPS) will measure the composition, distribution and energy of charged particles in the Mercurian magnetosphere. The Atmospheric and Surface Composition Spectrometer (MASCS) is sensitive to all wavelengths from infra-red to ultra-violet, and will measure the abundances of atmospheric gases as well as detecting minerals on the surface, while the Radio Science experiment (RS) will use the Doppler effect to measure slight changes in the velocity of the space-craft as it orbits Mercury. This will help in tracking down variations in the thickness of the crust, as well as possible dense areas below the surface of the same type as the lunar mascons.

The instruments were placed on a "science deck" facing Mercury, and most of them are fixed, so that the coverage of the surface is obtained by Messenger's actual motion over the planet. Remember, everything had to be as lightweight as possible. It was an ambitious programme, but there

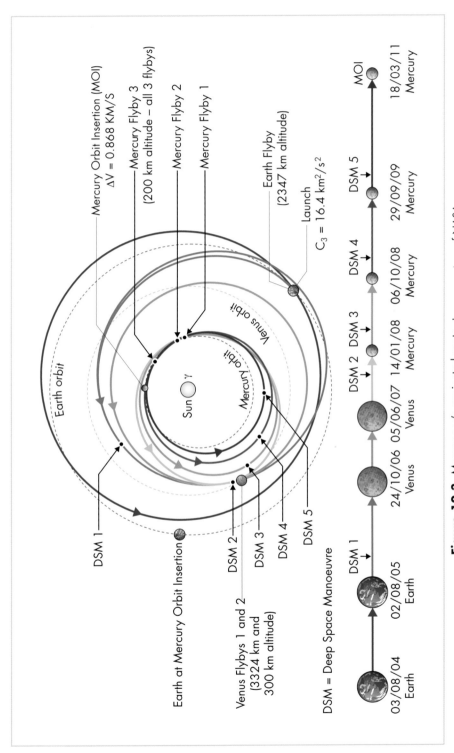

Figure 12.3. Messenger's projected route. Image courtesy of NASA.

seemed to be no reason why it should not be successfully carried through. There was no back-up mission (the accountants again!) so that everything depended upon this one spacecraft.

The launcher's limited power meant that the journey would have to be complicated; the planners thought wistfully about the Saturn-5 rockets used for the Voyagers, but Messenger had to make do with what was available. Without using the "gravity assist" technique, involving both Earth and Venus, reaching Mercury would have been wellnigh impossible, and there was a precedent, since Mariner 10 had by-passed Venus during its journey, and had sent back some useful information about that decidedly unwelcoming world. Messenger would make one fly-by of the Earth and two of Venus, followed by four fly-by encounters with Mercury itself before going into its final orbit. Deep-space manœuvres would be carried out in early 2006, late 2007, early and late 2008 and early 2010. If all went well, this would be the course of events:

Launch	2004 August 3
Earth fly-by	2005 August 1
Deep Space manœuvre	Early 2006
Venus fly-by	2006 October 24
Venus fly-by	2007 June 6
Deep Space manœuvre	Late 2007
Mercury fly-by	2008 January 15
Deep Space manœuvre	Early 2008
Mercury fly-by	2008 October 6
Deep Space manœuvre	Late 2008
Mercury fly-by	2009 September 30
Deep Space manœuvre	Early 2010
Mercury Orbit insertion	2011 March 18

Of course, this schedule may have to be modified, but we can only hope that nothing drastic goes wrong. There are no service stations in space.

Messenger will not be idle during its tortuous journey. In particular, data will be collected during both the passes of Venus, the first in 2006 at a range of 1600 miles, and the second in 2007 at 2670 miles. It is not really probable that much will be added to the existing data, but again one never knows, because Venus is anything but a static world. There is every reason to suppose that active volcanoes are erupting all the time, and Messenger will be able to make useful measurements, notably by checking on the amount of sulphur dioxide being sent into the atmosphere by the eruptions of volcanoes.

Sending a space-craft to Mars or the outer giants means giving the probe sufficient speed. With Messenger, the problem is in slowing it down as it reaches Mercury after a journey which has covered almost 5000 million miles, and has meant fifteen trips round the Sun. The use of retro-rockets is the only practicable method. Before orbit insertion, the velocity has to be reduced relative to Mercury; this is done by pointing the thruster forward and firing it for fifteen minutes. A shorter burst two to three days later should bring the speed down to the correct value, setting Messenger into a stable orbit round Mercury.

Initially the distance from the planet will range from 124 miles at the lowest point out to 9520 miles at the furthest, in a period of 12 hours. The inclination of 80 degrees means that the lowest point lies over the northern hemisphere, giving the best possible views of the Caloris Basin. Solar radiation pressure will cause a slow change in the orbit, but this can be corrected by regular bursts of power – for which Messenger has ample fuel reserve.

Other changes in the orbit will be caused by Mercury itself. The orbiting lunar probes led to the discovery of mascons (MASs: CONcentrations) below the circular seas, such as the Mare Imbrium, and some of the large walled plains, such as Grimaldi. The material below the crust in these regions is above average density, and will make an orbiting vehicle speed up as it approaches that part of its circuit. (It was originally suggested that a mascon might be a buried, iron-rich meteorite, but this does not seem to be

Figure 12.4. Artist's impression of Messenger nearing Mercury. Courtesy of NASA.

the case, and a mass of volcanic rock is a better solution.) Whether or not Mercury has similar mascons remains to be seen. The orbit has been planned so that Messenger will pass quickest over the hottest areas.

It is hoped to go on collecting data for one Earth year, which covers two Mercurian solar days, but probes have a habit of lasting well beyond their "sell-out" dates; the latest Mars robots, Spirit and Opportunity, give us a good example of this.

Such is the current state of play as I write these words (June 2006). The journey will take a long time, and one inconvenient point about this is that many of the original team will be retired or dead before the main results come in. Probes to Mercury were planned in the 1960s; nearly half a century will elapse between the pass of Mariner 10 and the Messenger encounter.[12] Sadly, nothing can be done about this.

Messenger is not the only probe to Mercury due in the reasonably near future. There is also BepiColombo, a joint venture by the European Space Agency (ESA) and the Japan Space Exploration (JAXA). (It was named in honour of Giuseppe ("Bepi") Colombo, of the University of Padua, who originally suggested using the gravity of Venus in order to help Mariner 10 on its way to Mercury; he was born in 1920 and died in 1984.)

BepiColombo will be launched in 2013, just as Messenger will be nearing the end of its career. The journey will take $3\frac{1}{2}$ years, ending up in orbit round Mercury in 2019 – at least, this is the current schedule; it may be modified. There will be two space-craft involved in the mission, the Mercury Planetary Orbiter (MPO) and the Mercury Magnetospheric Orbiter (MMO). A small lander, the Mercury Surface Element (MSE) may be added – if one can avoid the baleful glare of the accountants! Launching details have yet to be decided. The two orbiters alone could be sent up from Russia's Balkonur Cosmodrome in Kazakhstan, using a Soyuz-Fregat rocket; with the lander, the launching vehicle would more likely be an ESA Ariane-5, from Kourou in French Guiana.

Until we receive the first data from Messenger, we cannot even speculate about what BepiColombo will tell us, but no doubt the results will be fascinating. I am only sorry that I am unlikely to be around when they come in.

---

[12] In this respect, perhaps the unluckiest investigators are those who in 1960 used a powerful radio telescope to send a signal to hypothetical beings residing in the Hercules globular cluster. As the cluster is 17,000 light-years away, we may expect a reply about the year AD 35,960. Wait for it!

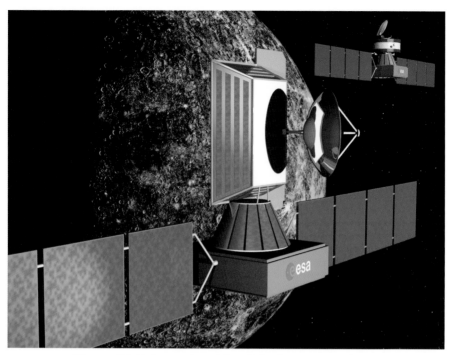

Figure 12.5. Impression of BepiColombo. Courtesy of ESA.

Life on Mercury?

Men have been to the Moon; unless we allow our politicians to drag us into further wars, men will go to Mars at some time during the present century.[13] But where else? Venus, with its searing heat, crushing atmospheric and acid-rich clouds, does not look attractive, and the satellites of Jupiter and Saturn are inconveniently far away. This leaves us with Mercury. What are the chances of life there?

There was a time when most planets were believed to be inhabited, and some past writers were enthusiastic. Among these was Emanuel Swedenborg, who in his younger days carried out some good scientific work. He was Swedish, and became well known all over Europe. He was born in 1688, and died in 1772.

Swedenborg's writings changed as he grew older. He explained that he was in constant touch with spirits and angels from other worlds, which of course gave him a head start over everybody else. All the planets were indeed inhabited, said Swedenborg, and he was given detailed information about these other beings. For example, some of the spirits of Mercury took the form of crystalline globes – in fact, they were balls – but they were well versed in abstract knowledge: "The spirits of Mercury imagine that they know so much, that it is almost impossible to know more" – a view shared by at least one current politician! They enjoy a moderate climate, without

---

[13] Note that I say "men". This may not be Politically Correct, but I must point out that the correct definition of the human race is *Homo* sapiens.

Figure 13.1.
Emanuel Swedenborg.

extremes of heat or cold, even though the Sun looks larger than it does from the Earth. Swedenborg had the answer to this. "It was given to me," he explained kindly, "to tell them that it was so provided by the Lord, that they might not be exposed to excessive heat from their greater proximity to the Sun, since heat does not arise from the Sun's nearness, but from the height and density of the atmosphere, as appears from the cold on high mountains even in hot climates." In fairness to Swedenborg, it has to be said that even today some people are puzzled by the fact that mountain-tops are somewhat chilly!

Many people agreed that most planets might be inhabited; William Herschel, discoverer of the planet Uranus and arguably the greatest of all telescopic observers, believed that the habitability of the Moon was "an absolute certainty", and was convinced that there were beings living in a cool region below the surface of the Sun. However, there were some dissentients, such as the English philosopher William Whewell (1794–1866). According to Whewell, we must be unique, because the existence of other races would affect Man's special relationship with God.

Against this, a well known astronomical author, Thomas Dick, wrote in 1837 that for Mercury, "the surface area is sufficient to support the present population of our globe. Hence it appears, that small as this planet may be considered when compared with others, and seldom as it is noticed by the vulgar eye, it in all probability holds a far more distinguished rank in the intellectual and social system under the moral government of God, than the terrestrial world of which we are so proud". We are almost back to Swedenborg's spirits.

Coming to later times, it is worth considering the views of Richard A. Proctor, who was not only a good cosmologist but also one of the first great popularizers of astronomy. He wrote many books, and in one of them – *Other Worlds than Ours*, published in 1878 – he says something about possible life on Mercury.

He knew that life of the kind we can understand needs air and water. He was ready to accept that Mercury had a reasonably dense atmosphere and probably some underground water, so that on these counts life could not be ruled out. However, there was the problem of the high temperature. Proctor suggested that the polar regions might be habitable, and that the two polar settlements might be able to build long tunnels below ground, so that they could pass to and fro without having to brave the torrid regions of the equator. (He did not explain how the subterranean tunnels were to be built in the first place; certainly it would require engineering skill beyond the capabilities of even Percival Lowell's canal-building Martians.) He went on to point out that gravity on Mercury is much weaker than on Earth; "hence the creatures which seem to us most unwieldy – the elephant, the hippopotamus and the rhinoceros, or even those vast monsters, the mammoth, the mastodon and the megatherium, which bore sway over our own globe in far-off eras – might emulate on Mercury the agility of the antelope or the greyhound".

I admit that I am intrigued by the idea of a hippopotamus emerging from a tunnel and leaping joyously around the cratered landscape. Sadly for this picture, we now know much more about Mercury than Proctor did.

Mercury has always attracted science-fiction writers, including some of the calibre of Isaac Asimov. Sometimes we are regaled with stories of entirely alien life-forms, generally hostile and who are often classed together as BEMs or Bug-Eyed Monsters. I do not propose to go down this road, because trying to argue with a BEM-enthusiast is pointless; it is rather like trying to eat tomato soup with a fork. But until a few decades ago, there was one possible loophole which might have made very lowly life possible.

Life can appear in very unexpected places (such as the hydrothermal vents on our ocean floors), and it was believed that there was one part of Mercury where temperature conditions were tolerable. As we have noted, Schiaparelli and Antoniadi maintained that there was a "twilight zone" on the surface, between the area of permanent day and the area of permanent night. If the rotation period had really been equal to the orbital period – 88 Earth days – this would have been quite valid, and I suppose it is just conceivable that very lowly organisms might have survived. But as we found out during the 1960s, the rotation period is not the same as the planet's "year". The twilight zone does not exist; it vanished – and the last chance of Mercurian life vanished with it.

14 A Trip to Mercury

There is no doubt that a manned flight to Mercury is absolutely out of the question at the present time. Quite apart from the hostile conditions on the surface, there is the problem of getting there. It would be vastly more difficult than a journey to Mars, partly because Mercury is much further away and partly because of the increased danger from radiation, which would be bad enough even on a Martian journey but which would be much worse when travelling closer to the Sun.

If we govern ourselves sensibly, there is no reason to doubt that these obstacles can be overcome one day – perhaps during the twenty-second century. We must hope for the best. Meanwhile, there is no harm in going on an imaginary trip there – so what would astronauts expect to see on arrival?

The first thing to notice would be the lower gravity as compared with Earth. The surface gravity is only 0.38 that of what we are used to feeling, so that a man weighing 12 stones at home would weigh only just over four and a half stones on Mercury. There would be no danger in this; after all, there have been six successful Moon landings, and the explorers have had no difficulties at all. Many people have watched the Apollo missions, and have seen the astronauts walking in what looks like slow motion, while the lunar rovers were driven quite happily across the landscape. Neil Armstrong once told me that reduced weight was "a very pleasant sensation".

In some ways the scene will be not too unlike that on the surface of the Moon, with craterlets very much in evidence. Presumably the landing site

Figure 14.1. The surface of Mercury. Illustrated by Paul Doherty.

chosen will not be near a hot pole at the time of perihelion! There are other points in common with the Moon. The atmosphere is absolutely negligible, so that shadows will be sharp and jet-black; there will be intense glare from the rocks, with a huge Sun looming down and heating the rocks below. The sky will of course be black. If the observers "dark-adapt" and then screen their eyes, they will be able to see stars at any time.

There is an interesting point to be made here. A question often asked is "Why did not the photographs taken from the Moon show stars?" The answer is, of course, that the photographs were not suited to this. To record stars a time exposure is needed, and this was not what was wanted for images of the lunar scene. (If you doubt this, go outdoors on a clear night and take a picture with an exposure of a second or two. I can assure you that no stars will be seen.)

Mercury must once have been a very active world, but all vulcanism ceased long ago, as with the Moon. Groundquakes are probably absent, and in any case cannot be strong enough to be dangerous. Neither can there be dust-storms or clouds. There is a magnetic field, far weaker than that of the Earth, but enough to make a very sensitive compass work. The main hazard will be radiation, and the onset of a solar storm will mean that the astronauts will have to take suitable precautions – possibly by retreating into an adequately-screened shelter.

The night sky will be familiar inasmuch as the constellations will look the same as they do at home, though they will be more obvious because of

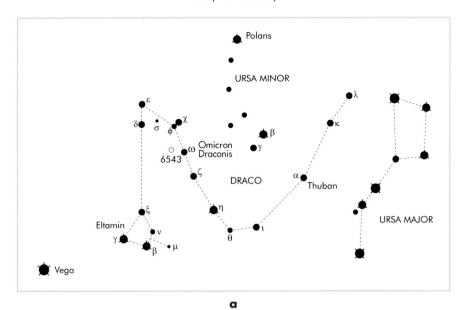

a

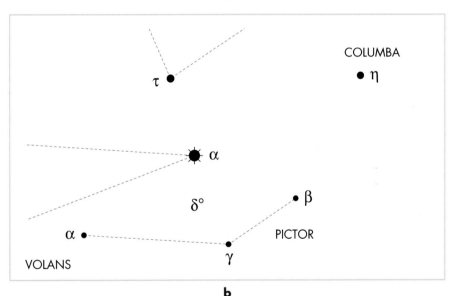

b

Figure 14.2. The pole stars of Mercury. **a** Omicron Draconis. **b** Alpha Pictoris. (The brilliant star is Canopus (Alpha Carinæ). The adjacent constellations are Columba (the Dove) and Volans (the Flying Fish).)

the lack of light pollution. However, the positions of the celestial poles will be different, and there will be different Pole Stars. The north celestial pole lies near an undistinguished orange star, Omicron Draconis, of magnitude 4.7; the south pole star, Alpha Pictoris, is rather brighter, at magnitude 3.3. Mercury's slow rotation means that the stars will shift across the sky in a very stately fashion. Of the planets, Venus would be magnificent, casting strong shadows, while the Earth would be as brilliant as Venus seems to us, and the Moon would be an easy naked-eye object. Mars would be conspicuous, and Jupiter and Saturn would be above the first magnitude for most of the time. Only the outer planets would be seen to less advantage than they are from Earth.

There would be no meteors; the atmosphere is far too thin to have any effect upon an incoming piece of cometary débris. On the other hand, there is every likelihood that the comets themselves will often be spectacular. Most "great comets" peak in brilliancy when they are near perihelion; seen from Earth they are then inconveniently close to the Sun in the sky, and are

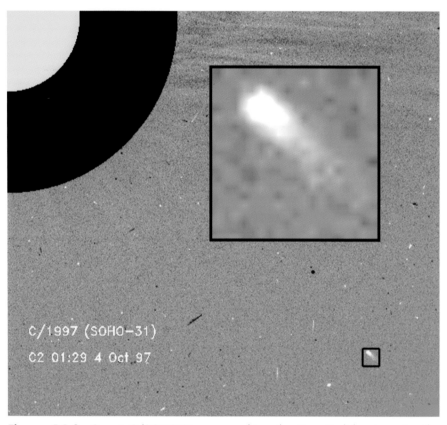

Figure 14.3. Comet C/SOHO-31, approaching the Sun. It did not survive the encounter. A Mercurian astronomer would have had a fine view of its last moments! Image obtained by the LASCO instrument, on the SOHO satellite, built and operated by the LASCO consortium of the Naval Research Laboratory (Washington D.C.), The Laboratory for Space Astronomy, Marseilles (France), The Max Planck Institute for Aeronomy, Lindau (Germany) and The Department of Space Research, Birmingham (UK). SOHO is a joint ESA/NASA mission of international cooperation.

drowned in the glare, but from airless Mercury they will be much easier to study. There are also many comets which actually fall into the Sun, and from Mercury observations of these cometary suicides will be fairly common.

Some asteroids also move in the inner reaches of the Solar System. We have tracked many of them; as yet we have found no "Vulcanoids", but they may well exist, and if so will no doubt make close approaches to Mercury (in fact, many of the craters on the surface may have been caused by Vulcanoid impacts). Auroræ require the presence of an atmosphere, and on Mercury must be ruled out, but not so for the Zodiacal Light, which is caused by diffuse material, spread along the main plane of the Solar System. From Mercury both it and the Gegenschein may be quite bright.

All this seems reasonably straightforward. Antoniadi, much the best of the pre-Space Age observers, had different ideas about the situation there, mainly because he believed in a region of permanent day, an area of permanent night, and a comparatively friendly twilight zone in between. He also claimed that the atmosphere was dense enough to support clouds, and that volcanic action could not be ruled out. Let us look back to what he wrote in 1932:[14]

A more-than-glacial breeze would be extremely punishing during the night, while a wind incomparably more scorching than the desert simoon would probably give rise to a spectacle of fuming dunes, which during the day would raise eddies of greyish dust which would cover the sky, and conceal even the Sun with sinister, all-obscuring clouds. Moreover, there would be literally no humidity, evaporation or condensation; no cirrus, stratus, cumulus, nimbus, waterspouts or mists, neither rain, ice, snow, dew nor white frost. Next, the absence of water-droplets and ice-needles in the air would mean no rainbows, haloes, parhelia or mock-suns; there would not even be coronæ; and in spite of the very great heat of the shifting Sun, it is doubtful whether mirage effects could transform the desert plains into optical lakes.

We have learnt a great deal since Antoniadi's time, and it is fair to say that Mercury seems even less welcoming than it did a few decades ago. Yet it is as intriguing as ever, and we may still hope that in the years to come men from Earth will be able to visit this curious little member of the Sun's family.

---

[14] Antoniadi, The Planet Mercury, Paris, 1932. Page 74 of my English translation, 1974.

15 Mercurian Base

Messenger is on its way to Mercury; BepiColombo and other missions are planned. So what lies in the future with regard to the exploration of the planet?

It is, to put it mildly, most unlikely that anyone reading this book will live long enough to see astronauts setting out for Mercury. Sooner or later it may happen – if it is considered worthwhile. Meantime, there is no harm in indulging in a little speculation.

The first observatories to be put there will of course be purely automatic, and will have to manage without human maintenance. They will be of great value. After all, Mercury is an excellent site for observing the Sun, finding and examining comets, searching for Vulcanoids and much else. There seem to be no insuperable difficulties here, and the first robotic stations may well be in place within the next few decades. But manned flight is another matter, and any pioneers will run risks that may make a journey to Mars look rather like a gentle stroll. It is not only the increased distance and the heat; we must also remember that close to the Sun the radiation danger increases enormously, and the crew of our space-craft will have to be permanently on watch for solar storms. Even if one occurs, what exactly can be done about it?

I feel that for travel to Mercury we are not looking at 2100, but more probably 2200 or 2300, and I am quite prepared to believe that it will never happen at all. It may be that a manned base could in the long run add little to what can be learned from a robotic observatory. If an attempt is made, it will not be a quick "there and back" flip, as with the Moon, or even an expedition comparable with a journey to Mars (we can probably forget all about Venus, where the conditions will be even worse than on Mercury, bearing in mind that Mercury has no crushing atmosphere or clouds rich in

Figure 15.1. A base on Mercury. Illustrated by Tony Wilmot.

sulphuric acid). In any case, we may have to wait until we can find ways of cutting down on the travel time. Ion propulsion is being developed (SMART-1, the latest lunar probe, uses it) but even so we need something much quicker.

Never mind; let us speculate. Assuming that a manned flight is made, what would a Mercurian base be like?

I think we must agree that a surface base is not practicable, unless perhaps in a polar region. Elsewhere on Mercury, the intense daytime heat would make conditions impossible. Setting up house in a polar crater, whose floor never receives any sunlight, does not seem so completely out of the question, despite the numbing cold. Yet it would be a decidedly gloomy place, and carrying out any really useful scientific researches would be difficult in the extreme. Perhaps excavating a site on the outer surface, or even utilizing a natural cave, would be better. This would mean setting all the recording instruments above ground, and thus available for servicing at any time. Monitoring the equipment could be done from the cave or the crater-floor, so that the astronauts would venture out only when there was no fear of an imminent solar storm. The danger from comets could be greater on Mercury than it is on Earth, and as yet we know nothing definite about the numbers of meteoroids in regions close to the Sun.

What would the Base itself be like? Wherever situated, it would have to be big enough to accommodate the whole expedition, and there would have to be a considerable number of people. There must be adequate medical provision, and efficient hospital equipment, for example. Social problems could not be neglected, because a tour of duty is bound to be protracted; the astronauts will have to be in each other's company for some time, and any personal tension under such circumstances would be a recipe for disaster. Whether the teams would be mixed or "men only" is a matter for debate!

Comfort inside the Base will be essential, and there must be ample provision for recreation, but it is hard to imagine that life there will be pleasant, and no doubt astronauts will be very glad when their tour of duty is over and they are on their way home.

Obviously the Base must be roomy, with suitable quarters for all who live there – and a proper kitchen, plus an experienced chef. Going for forays in the open will have to be very carefully timed, particularly with regard to events taking place in the Sun. It will be hard to cope with emergencies, and we cannot even think seriously about an expedition to Mercury until we know a great deal more about space travel than we do now. At least there should be no real communications problems. There will be times when direct transmissions are interrupted, when Mercury and the Earth are on opposite sides of the Sun, but even then it will be possible to use orbiting relay stations – as well as the bases on Mars, which will certainly have been established by then.

Figure 15.2. A base on Mercury at night. Illustrated by Tony Wilmot.

Most of these ideas may turn out to be very wide of the mark, and all we can do is to wait and see. Mercury is not in the forefront of our manned targets in the Solar System; it is well behind the Moon and Mars, and even possibly some of the satellites of Jupiter and Saturn, though ahead of Venus. Frankly, we do not yet know.

At least our ideas now are reasonably clear-cut. Mercury shines coyly down at us in the evening twilight or the glow of dawn, and it is beckoning to us. Unwelcoming – yes. Uninteresting – no!

Appendix 1

Data for Mercury

Distance from the Sun, miles: max 43,390,000 (0.308 a.u.)
      mean 36,000,000 (0.387 a.u.)
      min 28,590,000 (0.306 a.u.)

Orbital period: 87.969 days

Eccentricity: 0.206

Orbital inclination, degrees: 7.004

Orbital speed, mi/sec: max 36.7, mean 29.4, min 24.2.

Synodic period: 115.8776 days.

Apparent diameter from Earth, seconds of arc: max 12.0, min 4.5.

Equatorial diameter, miles: 3031

Oblateness: negligible

Mean surface temperature;   day 662 degrees F, 350 degrees C;
      night –274 degrees F, –170 degrees C.

Extremes of temperature, degrees C: day +427, night –183.

Mean diameter of Sun, as seen from Mercury: 1°22′40″.

Mass, Earth = 1: 0.055

Escape velocity, mi/sec: 2.7

Equatorial rotation velocity, mi/h: 6.77.

Mean density, water = 1: 5.44

Surface gravity, Earth = 1: 0.38

Volume, Earth = 1: 0.056

Axial inclination: –0.01

Reciprocal mass, Sun: 1:6,000,000

Inclination of equator to orbit: 0 degrees

Positions of celestial poles: North RA 18h 44m, dec +61.45 degrees
      South RA 06h 44m, dec –61.45 degrees.

(Continued on page 132)

Pole stars: North, Omicron Draconis (mag.4.7).
              South, Alpha Pictoris (mag.3.3).

Maximum apparent magnitude: −1.9.

Albedo: 0.06

Greatest elongation from the Sun, degrees: 28.3.

ATMOSPHERE (or, more properly, Exosphere):

Ground density, about 10^{-10} millibar.

Total mass, about 8 tons.

Constituents: potassium 31.7%
                sodium 24.9%
                atomic oxygen 9.5%
                argon 7.0%
                helium 5.9%
                molecular oxygen 5.6%
                nitrogen 5.2%
                carbon dioxide 3.6%
                water 3.4%?
                hydrogen 3.2%

Appendix 2
Phenomena of Mercury

ELONGATIONS, 2005–2012	
Eastern	
2005	12 Mar, 9 July, 3 Nov.
2006	24 Feb, 20 June, 17 Oct.
2007	7 Feb, 2 June, 29 Sept.
2008	22 Jan, 14 May, 11 Sept.
2009	4 Jan, 26 Apr, 24 Aug, 18 Dec.
2010	8 Apr, 7 Aug, 1 Dec.
2011	23 Mar, 20 July, 14 Nov.
2012	5 Mar, 4 July, 26 Oct.
Western	
2005	26 Apr, 23 Aug, 12 Dec.
2006	8 Apr, 7 Aug, 25 Nov.
2007	22 Mar, 20 July, 8 Nov.
2008	3 Mar, 1 July, 22 Oct.
2009	13 Feb, 13 July, 6 Oct.
2010	27 Jan, 26 May, 19 Sept.
2011	9 Jan, 7 May, 3 Sept, 23 Dec.
2012	18 Apr, 16 Aug, 4 Dec.

TRANSITS, 2006–2050.		
		Mid-transit (GMT)
2006	Nov 8	21.42
2016	May 9	14.59
2019	Nov 11	15.21
2032	Nov 13	08.55
2039	Nov 7	08.48
2040	May 7	14.26

Bibliography

ANTONIADI, E.M. (translated by Patrick Moore).
The Planet Mercury. Keith Reid, Shaldon 1974.

CROSS, C.A., and MOORE, P. Atlas of Mercury.
Mitchell Beazley, London 1977.

MURRAY, B. and BURGESS, E. Flight to Mercury.
University of Columbia Press 1977.

SANDNER, W. The Planet Mercury.
Faber and Faber, London 1963.

STROM, R.G. Mercury: the Elusive Planet.
Cambridge, Mass., 1987

STROM, R.G. and SPRAGUE, A.L. Mercury: the Iron Planet.
Springer-Verlag, London 2003.

VILAS, F.; CHAPMAN, C.R.; and MATTHEWS, M. (editors). Mercury.
University of Arizona Press, 1988.

NASA ATLAS OF MERCURY: Random House, 1988.

Index

137

Printed in Singapore